P9-EDX-000

Studies in Phenomenology
and the Human Sciences

STUDIES IN PHENOMENOLOGY AND THE HUMAN SCIENCES

edited by
John Sallis

HUMANITIES PRESS
ATLANTIC HIGHLANDS, N.J.

Reprinted from RESEARCH IN PHENOMENOLOGY 1979 Volume IX
Published in book form in the United States of America by Humanities Press, Inc. with permission of the editors.

Printed in the United States of America

Library of Congress in Publication Data
Main entry under title:

Studies in phenomenology and the human sciences.

1. Phenomenology—Addresses, essays, lectures. 2. Humanities—Addresses, essays, lectures. 3. Hermeneutics—Addresses, essays, lectures. I. Sallis, John, 1936-
B829.5.S78 142'.7 79-25653
ISBN: 0-391-01702-0

CONTENTS

Hermeneutics and Textuality: Questions Concerning Phenomenology

SAMUEL IJSSELING
University of Leuven

The most important contemporary French thinkers have all had to some degree a phenomenological education. This is obviously the case for J. Derrida whose first publications deal with Husserl. However, it is also true for J. Lacan and for M. Foucault and even to a certain extent for L. Althusser. Yet it is difficult to call these thinkers phenomenologists. Sometimes they are called by a rather problematic term—structuralists. The term is problematic, as all these thinkers reject structuralism as a characterization of their way of thinking. It is more correct to say that they are thinkers of discourse and discursivity, of text and textuality, of écriture and texture. It will later become evident that they are first of all readers or possibly listeners. They are concerned with the act of reading and writing, and of listening and talking. Consequently the style of their thinking and the thema of their work differ totally from those of phenomenology, in which perception and everything that it implies and presupposes is the central focal point. Merleau-Ponty published in the forties an important work entitled *Phénoménologie de la perception.* From the late fifties until the early seventies it would be difficult to find an important work or publication with the word perception or experience in the title, or even in the table of contents. However, words like *'lecture,' 'lire,' 'texte,' 'texture,' 'textualité,' 'intertextualité,' 'discours,' 'écriture,' 'signe,' 'signifiant,' 'signifié'* (and 'metaphor' and 'metonymy') occur very often.

The leading contemporary French philosophers have certainly had phenomenological training, but they have, as Levinas explicitly says of his own thought, used phenomenology to think against phenomenology. Furthermore, they have formulated radical questions concerning the basic ideas and structure of the phenomenological way of thinking. These questions on the one hand arise out of a radical consideration of the most fundamental notions and starting-points of phenomenology itself. Therefore a basic knowledge of phenomenology is necessary to understand what occupies these French thinkers. On the other hand these questions emerge out of a confrontation with many other tendencies in twentieth-century European thought.

First of all, I want to show schematically the fundamental notions of E. Husserl's phenomenology, which will serve as a background for contemporary French thought and which become critically questioned. At the same time however I would also like to consider other tendencies in European thought which have made it possible to raise these questions. Secondly I would like to give a short analysis of Husserl's basic ideas; this analysis will be somewhat tendentious, as it already shows the problematic of these fundamental ideas. And thirdly I will endeavour to say something about what occupies French philosophers, through particular attention to the act of reading. This will be an opportunity, although limited, to enter into the problem of hermeneutics and textuality.

FUNDAMENTAL NOTIONS OF PHENOMENOLOGY AND OTHER TENDENCIES OF CONTINENTAL THOUGHT

Perception is central for Husserl, both sensory perception with the material object (*Gegenstand*) given to us, and non-sensory perception with its so-called ideal object (*ideale Gegenstand*) given to us. It is important in an analysis of perception to take into account that the word perception points to what is perceived (what is present) as well as the act of perception itself (the making present). Perception on the one hand necessarily refers to a perceiving subject for whom the object can be object and what is present can be present, a consciousness for whom the given is given, for whom the appearance appears (phenomenon). On the other hand, perception necessarily refers to the givenness of that which is given or the presence of the present. Phenomenology as *logos* (discourse) of the phenomenon must then give a description, as adequately as possible, of what appears in and to consciousness. This description must be completely and always borne by consciousness, or by a fully aware and writing subject who gives an account of what has been said or written. Adequate description is a never completed and a never to be completed task. Knowledge, perception and description are temporal processes, and consequently time plays an essential role in all of this. We have mentioned the fundamental words in phenomenology: perception, consciousness, presence, subject, the being given, and the adequation of speech and writing with the present (or the given) and time (such adequation was at least, in principal, aspired to). These are the basic words, which, with their equivalents, carry and make possible, structure and lead the argument of Husserl. They continuously refer to and presuppose each other. They form the material of which the text of Husserl is woven. It is precisely this network of words that contemporary French thinkers question. Critical remarks and considerations are made concerning the role of perception in philosophy and science, the status of the subject or subjectivity, the structure of consciousness, the notion of presence or givenness, the nature, the function, the status of the act of speaking and writing and of the spoken and written word, the concept of time, and finally the network of relations and oppositions between these concepts in the texts of Husserl.

These questions and critical considerations are inspired by a radical reflection on the argument of phenomenology itself, as well as by other factors. We will consider a few of them here, albeit only schematically.

Of prime importance is the structural linguistics and semiology of F. de Saussure, in which the significance of linguistic signs is linked to the place they have within the framework of signs. This framework is a whole of differences, or of oppositions and relations. For philosophy the significance of this was first discovered by Merleau-Ponty. A sure turning point in French thinking occurs with Merleau-Ponty's lecture "Sur la phénoménologie du langage," which he presented at the International Phenomenological Colloquium in Brussels (1951). In this exposé a number of ambiguities and unsolved problems in Husserl were pointed out and explicit reference made to de Saussure.[1]

Next and of great importance is Heidegger's destruction of ontology. Destruction is a rather misleading term, because it deals in fact with a structural analysis of the prevailing ontology in which the existent, from

---

[1]*Signes*, Gallimard, Paris, 1960, p. 105-122.

the Greeks onward, was conceived as that which is present, and this presence, from Descartes onward, was interpreted as being present for a subject. Further there is Heidegger's approach to philosophy as *Sprachwerk,* as a tissue of words, in which he tries to analyze the production, the various shifts that have occurred, and the effects, i.e. all that issues from this.

Next there is the confrontation with Freudian psychoanalysis which emphasizes the importance of verbalization and not perception. The exchange of words between the analyst and the patient is according to Freud the only thing that can and may occur in analysis. What is important in this verbalization is not the adequacy but a strategy of desire. Here we also find the notion of unconsciousness which is precisely that which does not appear and yet is the condition of possibility for consciousness and for each form of presence or appearance. The "royal way" to unconsciousness for Freud is the dream which is always a related dream and is seen as a rebus or a text. The unconscious then appears to be that part of the discourse which is not accessible to man. In Freudian psychoanalysis the Ego (subject) is not an original instance but comes into being, and arises from identification by and with others. Subjectivity and identity are then always acquired subjectivity and identity.

Then there is Marx who maintained that subjectivity and consciousness are not originally given, only derivative and secondary, but who also believed that there is a clear distinction between the object of perception and the object of science. Finally there is Nietzsche—probably the most important thinker for contemporary French authors—for whom subject and object, the I and reality, are only fictions and interpretations which are supported by a hidden will to power, effects or products of a transmitted grammar and an actually existing spoken and written word. This spoken and written word does not strive for adequation with the so-called reality but is a strategic manoeuvre to create order. This manoeuvre is an aspect of the strategy of power, which although intentional is not subjective. What is important here is the order of the discourse itself; that order is a fragmented and old transmitted story that circulates among us.

Here we cannot linger on the renewed reflection concerning the status of human sciences, the importance of literary sciences (which give an instrument for the analysis of philosophical texts as literary products, including those of Husserl); and on the extremely important influence of surrealism on contemporary French thought. So far we have covered the basic notions of phenomenology and the background out of which these notions have been questioned. We will now try to develop this further.

THE PHENOMENOLOGY OF HUSSERL

The classical phenomenology of Husserl is not an unchanging and static system of philosophical assertions and methodological starting points. Husserl is first of all a searching and inquisitive thinker, *ein ewige Anfänger,* as he called himself. Undoubtedly there is a continuity from the *Logische Untersuchungen* (1900-1901) to the *Krisis der Europäischen Wissenschaften und die transzendentale Phänomenologie* (1936) but one must not forget the significant changes and shifts in this work. Some of these changes, for example the shift to a transcendental problematic, caused many of his students to distance themselves from Husserl and to go their own way.

Discounting these developments, and consequently oversimplifying, we can say that Husserl searches for a grounding of philosophy as *strenge Wissenschaft,* i.e. a philosophy that actually deals with reality. This grounding, according to Husserl, can only be found in the things themselves—hence his dictum: *zurück zu den Sachen selbst.* The foundation for philosophy is reality itself as it presents itself, as it appears, as it is given or manifests itself. This sounds reasonable and seems to be very self-evident. But Husserl claims that the thing itself, which is the issue for him, is consciousness. Phenomenology is the science of consciousness. Why consciousness? First of all, it should be made clear that consciousness is not something immanent, but essentially intentional. Consciousness is always consciousness of. This means that where it is said that consciousness is the thing itself or the foundation of all knowledge, one means consciousness with all that appears in and to it, or is present to it. More technically stated: consciousness with its noetic-noematic structure and its objects, which constitute themselves as objects in consciousness, is the only absolute being.

It is in the above light that the two interwoven methodological starting points of phenomenology must be understood: the *phenomenological reduction,* and what Husserl calls the *Prinzip aller Prinzipien* or the phenomenological principle.

The phenomenological reduction consists in turning from the things and turning toward consciousness in which or for which the things are present or given, in which the things as things, and the objects as objects, constitute themselves. This reduction is coupled with the *epochê,* which means bracketing the world. The *epochê* is neither a negation of the world nor a Cartesian doubt of the reality of the world. On the contrary it is a return to and a radical limitation of what appears in, to, and for consciousness. One must radically limit himself to what he actually sees or perceives, what is really given or evident, thereby explicitly refraining from every inclination toward any form of theory or construction. Later Husserl will say that the most important aspect of the epochê is not the bracketing of the world but the interpretations thereof, or of that which has already been said of the world. When restricting oneself to that which appears to or for consciousness, i.e., the phenomenon, there is no apparent reason for claiming that the phenomena are immanent. On the contrary, I do not see trees in my consciousness but there in the world, and I do not hear songs sung in my head but there on the stage. Objects which are *there* appear to consciousness.

The phenomenological principle that is closely tied to this states that every original given observation is an authoritative source of knowledge. Everything that presents itself in primordial form is simply to be accepted as it gives itself, but within the limits in which it presents itself. Therefore phenomenology is also descriptive psychology. It gives a precise description of what presents itself to consciousness, i.e. of the given as given, of the present as present, or of the perceived and the act of perceiving.

It is important to see that there are different forms of perception, and therefore different forms of being present or given. Accordingly Husserl distinguishes between ordinary sense observation and categorical observation to which the two forms of presence correspond. An example will illustrate the above. When I see a house and a tree and say that the house is mine, I am perceiving a house and a tree which are present for me. However, the "and" (conjunction) is also present, in one way or another, as the result of an act (conjunction). This is also true for the "this" ("thisness"), the "is" (being), the identity, and the relation to myself as owner, etc. Of course these are all present in a different way than the house, the tree, etc., but they are present for me as what Husserl calls *ideale Gegenstand,* and this within a *kategoriale Anschauung*. Certainly the whole matter is more complicated than presented here, but what is most important is that there are various forms of presence. Thus the past is also present in memory, or as sedimentation. The future is present in anticipatory expectation or as possibility. Also what is not factually present can, for example, be made present in the imagination or fantasy.

In the presence of that which is present time plays a fundamental role. Here we come upon an essential, yet problematic point in Husserlian phenomenology. In traditional philosophy time is not seriously considered. Time is eventually dealt with in philosophy of nature but otherwise plays no role except as a menace for real knowledge. Metaphysics would be an attempt to exceed time through thought or to arrive at eternal and immutable being. For Husserl it is totally different. Here time is no longer a menace for the presence of that which is present but a condition of possibility for the phenomenon as phenomenon. The real is for Husserl that which persists in a temporal, never to be completed perception or knowledge process. That which does not persist in this process is *only* appearance, or phantom, and not a real appearance. Later Heidegger will radicalize this and claim that temporality is a transcendental condition of possibility for *Dasein,* for the presence of the present, and for being itself. That which is present is something which comes into presence, and being is the event of coming into presence. Therefore, for him the real and primordial mode of time is the future. What is present is always coming. For Husserl the now or the present is and remains the primary mode of time; this becomes clear in his analysis of time-consciousness in which *retention* (the retention of the past) and *protention* (the anticipation of the future) are thought and thematized out of the present. The past is still present, and the future is already present.

On this point Husserl remained under the influence of traditional metaphysics which only knew one actual and real mode of time—the *praesens* (present tense) or possibly the *perfectum* (present perfect tense). The latter is obviously the case with Hegel for whom the central issue is presence which has always, in principle, already come to perfection. The fact that traditional metaphysics sees the now as the only real mode of time corresponds to its limited conception of reality or of being. Reality is put on the same footing as what is here and now present. The past is not (present) anymore, and the future is not (present) yet. That Husserl sees the *praesens* as the real mode of time is consistent with the phenomenological reduction which, as we have seen, consists in a radical limitation of what is actually present for consciousness; it is consistent also with the central place he gives to perception, to objectifying knowledge and theoretical acts, thereby leaving little or no place for practical dealing with being and still less for enjoyment and desire.

Husserl's emphasis on the present has enormous consequences for his concept of reality and his concept of language. For him reality is what is present for consciousness. A word or expression "points to," "refers to," or is "a sign of" a reality present at least in principle or a present relation in reality. We have already seen that for Heidegger the *futurum* (future tense) is the actual and the primary mode of time. Consequently reality for him is not what is present, but what is coming and possible; reality is not that which can be controlled or made objective but that to which one must freely open himself without possessing it in one's own power. He does not see the word as a sign of something but as the achieving of something: the coming into presence of what is present. To speak is not of the order of *Zeichen* (sign) but of *Zeigen* (showing, to make or let come into unconcealment). *Sagen ist zeigen.* Both Derrida and Levinas have shown that according to Heidegger this coming to be and this future, and thereby also language and speech, are still thought against the background of the now and the present. For people like J. Lacan—and this is crucial for understanding contemporary French thought—the human and primary mode of time is the *future antérieur* or *futurum exactum* (future perfect tense) (I shall have done that tomorrow). This is the mode of *non-presence* which is presented, feared, or desired as being present. For Levinas, and also for Derrida and Foucault that which is present is continuously retreating. It escapes from the discourse while the discourse imposes itself as the true reality. Presence is either that for which one hopes, or that which one fears will occur, while behaving as if it is already there; or like that which is lost for good with only a trace left behind and which one tries continuously to restore or find again. The latter occurs by means of signs and symbols. These do not refer to something that is present but to other signs and symbols, as words in a dictionary refer to other words. What is meant is only what is not present, and the world we live in is a framework of meanings. Reality is of the order of discourse or of the order of speaking and writing and of listening and reading. The most important question one then has to ask is: what precisely happens when one speaks, writes, listens and reads?

Before answering this question directly, we first have to say something about the problem of the transcendental in Husserl. According to Husserl, phenomenology would be not only descriptive psychology but also transcendental philosophy. What does this mean? According to Husserl, philosophy must not lose itself in various profound and metaphysical speculations, but it also must not try to compete with the positive sciences. It has to seek the transcendental. This signifies and the I that is signified. The Lebanese linguist E. Benveniste, who has worked for a long time in Paris and whose importance for possible, (from within and beforehand, a priori) and bears reality as present reality and perception as perception. Transcendental phenomenology tries to discover the most elementary structures, which we, here and now, are always already departing from, and which lead and make possible our perceiving and knowing, speaking and thinking, remembering and expecting. The search for the transcendental is not therefore for something outside the human reach, nor the search for an object within that reach. It is a search for the reach itself to which we are always and already accustomed, and that we continuously presuppose. This reach is the light within which we see what we see, and through which this seeing becomes possible. According to Husserl, the transcendental cannot be found in psychic structures as taught by psychologism nor in social structures as "sociologism" claims. Transcendental philosophy is something different from empirical psychology or positive sociology. Neither can it be reduced to historically realized structures, as historicism claims. The transcendental can only be found in the thing itself, in the intentional consciousness, which is the only absolute thing and cannot be reduced to something else. In connection with this, Husserl speaks about the transcendental subjectivity or the transcendental ego. What is present is present to a consciousness, and without this consciousness it is absurd to talk of presence. An object is always an object for a subject, and the given is always the given for something called the I.

Transcendental subjectivity or consciousness has been frequently objected to in the past from many different viewpoints. To escape from these difficulties, Heidegger has preferred the term *Dasein.* Consciousness and subjectivity are not original but deduced modes of *Dasein,* and therefore one can hardly maintain that they would be transcendental. Heidegger seeks the transcendental in time, in the temporality of Dasein, which functions as the condition of possibility or the presence of what is present. Following J. Hyppolite, one begins to speak of a *"champ transcendental sans sujet,"* a field that is not of the order of consciousness or of the I but one that would enclose and contain the conditions of possibility for subjectivity and consciousness. It would be the framework of oppositions and relations within which subject and consciousness come into being. Language or the order of discourse would be part of this transcendental field.

THE ORDER OF DISCOURSE

It is indeed language that causes the major difficulties concerning transcendental subjectivity. Husserl himself observed in his *Logische Untersuchungen* (II, 1, p. 82) that the I is not only an instance of thinking or consciousness but also of speaking. The I signifies itself in the act of speaking and this causes a remarkable duplication of the I that signifies and the I that is signified. The Lebanese linguist E. Benveniste, who has worked for a long time in Paris and whose importance for philosophy should not be underestimated, claims explicitly that the I comes into being with that signifying act. Language is the condition of possibility for subjectivity. It is only in and through language that one constitutes himself as subject (*Problèmes de linguistique générale* I, (p. 259-263). When one speaks about the constitution of the I or the subject through language or when one says that the I is an effect of the act of speaking, one can no longer claim that the I and subjectivity are original instances.

One of the problems encountered here is the very vague and ambiguous character of terms like 'language,' 'word,' 'speaking,' 'speech,' 'discourse,' etc., a fact not always sufficiently considered by the science and the philosophy of language. By the word 'language' one can mean the formal structure of the linguistic system; but one can also mean a certain given languge as, for example, English. One can also mean the act of speaking and writing (speaking and writing are not at all the same) but also an existing whole of spoken or written words, of expressions and texts, of stories and writings that are circulating. One can mean by that a system of signs which would have the function of signifying things and thoughts and of making possible communication among people, hence a usable instrument. However, 'language' can also be a kind of material or tissue out of which an expression or text is built up like an object (e.g. a fabric or a piece of art is built up of wool or wood). Moreover, this material may not be simply reduced to sounds or lines on a paper, because what is at stake is meaningful sounds and letters. One can still mean much more with terms like 'language' and 'word,' as, for instance, in expressions such as 'language games' and 'play on words,' the 'power of the word,' etc. Some confusion exists in this field, and it is not always clear what science or philosophy of language precisely covers.

What does Husserl say about language? For him the problem of perception and presence is central. His remarks on language are rather incidental, often a bit enigmatic and mostly underdeveloped. They are important, however, because in these remarks the problematic character of phenomenology's basic intentions come clearly to the fore.

In the *Logische Untersuchungen* (1900-1901) the word is seen as a sign and expression of a reality which is present for and in consciousness. The word or the expression would be the product of the thinking and conscious subject. Language would be constituted by consciousness in a sovereign and autonomous way, and thus at least in principle be completely clear. In the *Formale und transzendentale Logik* (1928-1929) Husserl goes a step further and understands language as body, as the embodiment of the thought. With this—although not explicitly thematized—the problem of opacity or the not completely transparent being of language is also introduced. It is not surprising that Merleau-Ponty, for whom the body as subject or the subject as body is central, takes this statement as point of departure. In the *Krisis der Europäischen Wissenschaften* (1936) Husserl speaks about the inseparable intertwining of man, world, and language. Man essentially lives in a society of language, and this is the horizon within which every experience can and must realize itself. Language becomes here the intersubjective possibility and limitation of experience. From this thesis, one passes easily to a conception of language as the condition of possibility for presence as presence, for consciousness and subjectivity, for identity and identification, difference and differentiation, objectivity and objectification. When this is taken seriously, all the basic principles of phenomenology begin to shift. Later, from the Jewish tradition, in which the narrative element is central, E. Levinas will say that no identity or differentiation is possible without a network of stories which is transmitted and which is circulating.

In *Vom Ursprung der Geometrie* (among the *Krisis* texts)[2] language and especially the written records or the written word finally becomes a necessary condition of possibility for science and therefore for our culture which is based upon the sciences. The *ideale Gegenstände* without which science could not exist, presuppose the written word. In this connection we can say that no formal logic is possible without pencil and paper or without blackboard and chalk. And, of course, it is not at all accidental that the first publication of J. Derrida, for whom writing is extremely important, is a French translation with extensive commentary on Husserl's text (*L'origine de la géométrie,* Paris, PUF, 1962).

The problems that emerge in relation to Husserl's concept of language are the most important points of approach for contemporary French philosophers and allow them to pose certain critical questions to phenomenology. Language presents the greatest difficulties for and the strongest resistence to the phenomenological reduction. Is language a product of the constituting consciousness and of the transcendental subject, or is language a condition of possibility for consciousness and for

[2]Beilage III, Husserliana VI.

the constitution of the subject? Are speech and writing or discourse and text of the same order as consciousness and the subject, as Husserl claims, or do they form an original and autonomous order with their own productivity and effectivity and with their own intentionality and operationality? This applies to speech and writing and to discourse and text in general, as well as to the speech and writing or discourse and text of Husserl himself.

Perhaps E. Fink, one of the most important co-workers of the older Husserl, adumbrated this problem when he spoke of "operational concepts" in phenomenology. Operational concepts are the concepts which Husserl needs to build his philosophy yet which are not explicitly thematized and thus do not belong to the order of consciousness. They are operational, that is to say, they function in the writing of Husserl and make this work possible. Perhaps it is better to speak of operational words rather than concepts. Concepts refer to comprehension and suggest an inner activity. In fact, Husserl's argument is supported by a framework of material words which make his argument possible and give a direction to it. Husserl makes use not only of words which he himself has chosen but of a number of words which are given to him and are active in his work. Thus another reading of Husserl becomes possible. One can begin to ask what is at work in the text of Husserl. How does this text come into being, and what does it accomplish?

Furthermore, one can wonder whether the word is a sign of present reality and refers to something that is somehow present or whether it presupposes absence and refers only to the other signs. Isn't the actual reality in which we normally live a world of words, of texts and stories, of sayings and expressions which are interwoven and thus presuppose and make each other possible. According to contemporary French philosophers the world in which we carry out our existence is indeed the world of discourse and of non-presence. This does not mean that there is an identity of reality and discourse because there is and remains an essential and irresolvable difference between discourse and reality. This difference makes possible the continuation of discourse and the production of new discourse.

Reality belongs to the order of discourse. Such a statement undoubtedly sounds curious, perhaps awakes irritation, and even meets with resistance. There are two considerations that can remove some of the strangeness and irritation and remove the resistance. Descartes tries to think away the whole of reality in order to discover that the only thing which cannot be thought away is thinking itself. The *cogito* is beyond doubt. This was very important for Husserl, because for him the *cogito* is the only absolute being; but he understands this *cogito* as intentional. One can make a similar movement of thought concerning language or the word. If one tries to think away every form of language and verbalization, nothing of what keeps us human beings busy seems to be left. This text would not be here, nor philosophy or science; there would be no literature, no administration of justice, no law, no politics, no schooling, no religion, no tradition, no institutions that support and are supported by them, and finally no human world. Perhaps there would be no God, because God is for us always a proclaimed God. Not one of us has ever seen God, we have only heard him spoken of. The second consideration is of a more empirical nature. When one considers seriously what we people usually talk about, and are occupied with, one realizes that it is first of all a world of words. We talk about what we have read and heard, about books and lectures, about literature and science, about the bible or other kinds of texts, about persons and events, which we know through stories, or about the interpretation and news of the world. This is especially the case for philosophers who of course continuously work with texts because philosophy consists first of all of texts. Even when one is doing phenomenology one is first of all occupied with the writings of the phenomenologists. However, this is also the case for the ordinary man who talks about the things he knows; almost everything he knows he knows from hearsay, and without this knowing there would be very little for him to do and experience.

The world in which we carry out our existence, is a world of words and is of the order of discourse. These words and this discourse form a reality all its own and to a certain extent an autonomous order. This does not mean that there would be an identity between what is usually called reality and the order of discourse. What is means is that we take our reality and identity, our place in the world and our position towards ourselves and others, from this special and material reality or the world of texts and from the autonomous order of the discourse. As a consequence our being is always acquired being, and our identity and reality a broken identity and reality. It is also on the basis of this special order of the word that such a thing as a meaningful present reality is possible. This presence is always permeated with a fundamental and irremovable absence, which is itself a condition for meaningfulness.

It is too simple to reduce this special reality of the words and this special order of the discourse to a framework of signs that would refer to the thoughts and presentations, aims and wishes of a speaking or writing subject. It is also too simple to reduce these to a system of signs that would refer to a present reality. Subjectivity and presence are more the effects of speaking and writing and not original instances. Words first of all refer to other words, every text refers to a framework of other texts, and each fragment of the discourse refers to other fragments. As Montaigne remarked, it is a fact that there are more books written about books than about any other subject. Of course, this referring can take very different forms. Sometimes it is an implicit or explicit quotation or the taking over of a formulation or a topic from them. It can be an acceptance or a refutation, further summarizing and developing what is actually said or written, clarifying and correcting it, rejecting and acknowledging it as authority, taking a lead from it (as in the *Inferno* where Dante took Virgil as his lead) or pushing aside (as Nietzsche was pushed aside in the work of Freud). It can even be a relation of simple copying and mechanical reproduction, etc. Concerning this framework of references one also talks about intertextuality instead of intersubjectivity.

Of course, all this has enormous consequences. We noticed already that contemporary French thinkers are first of all readers or possibly listeners. They deal with all kind of texts or with discourse. Their reading and listening is a very specific and unique occupation. We limit ourselves here to the act of reading. *Mutatis mutandis* one can almost say the same about the act of listening.

Schematically and therefore rather simply spoken, one can say that there are at least three possible forms of reading. One can read with an eye on the truth, which means asking oneself whether what is written is true. This is a typical philosophical reading and it is also Husserl's way of reading. Undoubtedly, Husserl has read much but he was not a real reader, as was Heidegger for example. What is of primary importance for Husserl is perception and the reality behind the text or the thinking and the thought about the reality that are expressed in the texts. Only what is said is important and whether this fits with reality. The verbalization itself is not important. It is only the sign of something else, the expression of the thought. Such a reading presupposes a concept of truth and language that is not at all self-evident. Moreover, it is striking that Husserl could not think without writing; this has resulted in the enormous quantity of manuscripts that Husserl himself kept very carefully and that are now kept in the Husserl-Archives in Leuven. But Husserl never thematized the act of writing itself.

One can also read with an eye on the sense or meaning: the so-called hermeneutical reading. Hermeneutics which is closely connected with phenomenology, at least in the Heideggerian version, asks for the (possibly hidden) sense or meaning of the text. One starts from an actually given text and searches for the conscious or the unconscious, expressed or unexpressed intentions of the author (or of the speaker), that is, of the subject as origin; and one looks for the real message that is enclosed in, under or behind the text. Also here, the verbalization as such or the materiality of the text is not of essential importance. The main concern in this reading is with saying in another and possibly better way that the author says or what is said in the text. It is because of this view that the later Heidegger has given up hermeneutics or at least formulated it in another way. In his confrontation with the poets he came to the discovery that it is absurd to think that one can say in different and better ways what the poet says. One has to take seriously what is written literally. What exactly is meant here by "to take seriously" and "literally" is not completely clear. In the basic rule of hermeneutics that says that a part of a text can only be understood from the whole and vice versa, one already finds a first indication of the framework of references that is at work in every text. H.G. Gadamer has added a very important dimension to the hermeneutic problem by speaking of a "Wirkungsgeschichte." The effects of a text which lie outside the field of the intentions of the author are important. The text itself is at work. With this, one arrives at the field of rhetoric that, besides the composition or production of a text, deals with what a text can operate upon and accomplish and what is needed for that. Thus it is not accidental that one can perceive in Gadamer a growing attention to rhetoric.

There is, then, a third possible form of reading: the rhetorical. Attention is here primarily on the materiality of the text or the textuality. Questions are raised concerning composition, that is to say, the structure and the production of the text, and concerning the actual and possible, direct and indirect, intended and not intended effects of the text. It is not the question of truth or sense that is central here but the question of how a text comes into being and what it brings about. The coming into being is not immediately coupled with a producing subject that, as author, would be the source and the origin of the text. The author, as author, is more a product or effect of the act of writing and of an act of ascription or attribution. Every text is only possible on the basis of an already existing whole of texts and is also necessarily woven in a framework of other texts. In that framework of texts there is a certain hierarchy. Some texts are dominating and others are dominated. In one way or another these texts are at work in the coming into being of the text, as the texts of Hume are at work in the work of Kant or the texts of Ricardo and Smith in the work of Marx. The coming into being of a text is furthermore continually accompanied by all kinds of systems of inclusion and exclusion, by all sorts of normalizing rules, and by different procedures concerning orthography, and style, thematic and composition, and disemination and publication. Primarily what a text brings about is the creation of the possibility of the production of new and different texts. Effects of the text also include especially the dividing and assigning of the different positions of the human being with regard to himself, the other, and the world, and keeping up and protecting or changing and subverting the framework of interrelations between human beings and between man and his world. This field of positions and relations exist on the basis of and thanks to an entire complex of texts.

One can approach a text by focusing on how it comes into being and what it brings about. Such an approach is perhaps far from every form of phenomenology. Perhaps! For Husserl, the phenomenon was what is present in and for the intentional consciousness. He has given an extremely scrupulous analysis of this phenomenon. In the previously discussed way of approaching texts, the text becomes the phenomenon, the matter itself. An equally scrupulous analysis of what happens in and around texts should be given. An important instrument for such an analysis is given to us by classical rhetoric, which may not be reduced to a study of metaphor and metonymy but in which the power by and of the word is central.

The Human Experience of Time and Narrative

PAUL RICOEUR
University of Paris

My aim in this lecture is to bring together two problematics that are not usually connected: the epistemology of the narrative function and the phenomenology of time experience.

On the one hand, the epistemology of narrative, whether it takes narrative in the sense of history-writing or story-telling, scarcely questions the concept of time which is implicit in narrative activity. It takes it for granted that narratives occur *in* time, i.e. within a given temporal framework, and it uncritically identifies this given temporal framework with the ordinary representation of time as a linear succession of abstract "nows."

On the other hand, the phenomenology of time-experience usually overlooks the fact that narrative activity, in history and in fiction, provides a priviledged access to the way we articulate our experience of time.

The main thesis of this paper will be that narrativity and temporality are as closely linked as is to a "language-game," in Wittgenstein's terms, corresponding "form of life." Or, to put it in different terms, narrativity is the mode of discourse through which the mode of being which we call temporality, or temporal being, is brought to language.

I shall first give a rough outline of the temporal problematics from the point of view of the ambiguities and paradoxes which may receive a specific clarification from our narrative activity. Then I shall turn to the narrative activity itself from the point of view of its temporal structures and inquire to what extent these temporal structures constitute an answer to the ambiguities and paradoxes of our ordinary experience of time.

I. THE PROBLEMATICS OF TIME

I have taken as evidence of the specific opaqueness of the human experience of time the eleventh book of Augustine's *Confessions* and Heidegger's *Sein und Zeit,* section II. In the context of this lecture, the convergence of their analyses will be held as more relevant than their obvious differences. I have selected three problems which seemed to me appropriate to further inquiry.

The first concerns the specificity of the human experience of time compared to the ordinary representation of time as a line linking together mathematical points. According to this representation, time is constituted merely by relations of simultaneity and of succession between abstract "nows," and by the distinction between extreme end points and the intervals between them. These two sets of relationships are sufficient for defining the *time when* something happens, for deciding what came *earlier* or *later,* and *how long* a certain state of affairs might last. But the deficiency of this representation of time is that it takes into account neither the centrality of the present as an *actual* now, nor the primacy of the future as the main orientation of human desire, nor the fundamental capacity of recollecting the past in the present.

The drastic move made by Augustine and by Heidegger was to say that there is no past, no present, and no future in any substantive sense, but rather a dialectic of intentionalities which Augustine referred to as memory, attention and expectation. For that purpose, he assumed the paradox of a threefold present, a present about the future, a present about the past, and a present about the present.

Heidegger makes the same move when he substitutes for the substantive terms of "future," "past" and "present" three modalities of what he calls "Care," which connect cognitive, practical and emotional components within one and the same ontological structure. I shall not consider at this stage the shift of emphasis thanks to which the present is deprived by Heidegger of its priority for the sake of the future. I shall focus rather on the phenomenology common to both Augustine and

Heidegger in order to emphasize the conflict between an experience of time rooted in the dialectic of Care and its representation in the terms of a linear succession of abstract "nows." This problem constitutes an enigma to the extent that the recovery of the genuine constitution of time does not seem to be able to abolish the representation of time as linear succession of nows in spite of the radical misunderstanding of the transcendal constitution of time that it generates. This resistance of the ill-famed reference of time must have some right of its own.

The second problem arises from the very paradox that both Augustine and Heidegger—albeit in very different ways—were compelled to introduce the above-mentioned opposition into our experience of time. For Augustine, the present of man is not the eternal present of God; for that reason, it is not a harmonious interplay of intentions—between memory, expectation and attention—but a discordant experience which Augustine called in a very appropriate way *distentio,* which means both extension and distraction. Let us reserve for the third paradox the quantitative aspects of *distentio* as extension and focus on the dialectic of intention and distraction. Intention is what prevails when, for example, in reciting a poem, we hold together the whole of the poem, in spite of the fact that a part of it is still ahead of us and another part has already sunk into the past and thus only a phase of the work is present.

Distraction is what prevails when we are torn between the fascination with the past in regret, remorse or nostalgia; the passionate expectation of the future in fear, desire, despair or hope; and the fraility of the fleeting present. *Human* time is the dialectic of intention and distraction, and we have no speculative means of overcoming it. Heidegger meets a similar enigma when he distinguishes between *authenticity* and *inauthenticity* in our relating to time. But inauthenticity, which reminds us of distraction, is not extrinsic to the purpose of an Analytic of *Dasein.* On the contrary, inauthenticity has its own existential claim which is that of *every-day life.* And since the description of everyday life belongs in an organic way to the analytic, we must introduce into our account of temporality the dialectic of authenticity and inauthenticity. To this drastic move we owe the most remarkable attempt to organize the phenomenology of time in terms of several *levels of radicality.* Accordingly, three such levels are distinguished. The most radical one is temporality properly so-called. It is characterized by the primacy of the future in the dialectic between the three temporal intentionalities and above all by the finite structure of time arising from the recognition of the centrality of death, or, more exactly, of *being-toward-death.* (The resoluteness with which we face our own being-towards-death, in the most intimate structure of Care, provides the criterion of authenticity for all of our temporal experience). We move in the direction of the inauthentic pole when we proceed from *temporality* to *historicity.* Historicity in its technical sense, refers first to our way of "becoming" between birth and death. The *stretching along* of life is thus more emphasized than the *wholeness* provided to life by its mortal termination. In this *stretching along* we may recognize Augustine's *distentio.* But this *distentio* is preserved from sheer dispersion thanks to *Dasein's* capacity to recapitulate—to *repeat,* to *retrieve*—our inherited potentialities within the projective dimension of Care. This *Wiederholung* is the counterpart of the stretching along of life. It witnesses to the continuity in the process of deriving historicity from temporality. We shall show at the end of this lecture the tremendous relevance of *Wiederholung* which surfaces in any attempt to ground historical *and* fictional narratives in a common temporal structure. We may now reach the pole of inauthenticity, in the sense of the prevailing of everyday life structures over and against those of temporality ruled by being towards death and of historicity ruled by repetition. The temporal structure corresponding to this stage is called *within-timeness,* because at that level, time is held as that "in" which events occur. We get closer to the linear representation of time, but the important claim of Heidegger is that before any *leveling* of *within-timeness* for the sake of the linear representation of time, its structure may still be referred to an analytic of *Dasein. Within-timeness* is still a feature of Care, but of Care as falling prey to its own objects, the subsisting and manipulatable things of one's concern. It is in this state of *thrownness* that the present becomes the predominant category. (But, even then the "now" is not the abstract "now" of the linear representation; it is the "now that" of human initiatives. It is al o for within-timeness that we make use of calculations and measurement. But we measure time because we *reckon with* time, and *reckoning with* time is part of the character of care as mundane *concern.* Within-timeness implies also public time, but this time is public not because it is *neutral,* indifferent to the distinction between things and men, but because human action is common action and because *Mitsein*—being together—may always and has always been reduced to anonymity. Even then, the anonymity of common time is not the abstractness of linear time according to the popular representation).

Such is the Heideggerian reading of Augustine's dialectic of *intention* and *distention*. But there is no speculative way to overcome this dialectic which remains unsolved and open.

A third and ultimate paradox arises from this very dialectic. It appears that *extension* is not the result of some fall, the sin of abstraction, but is a constitutive trait of the most radical temporality. We can trace it back from the mere extension of within-timeness to the *stretching-along* of historicity and finally the *ausser sich* of the three ecstasies of time. *Distentio,* then, is not merely a kind of disease of time-consciousness, but the *extension* which is dialectically connected to the intentionality of consciousness. It can ultimately be ascribed to the radical passivity pertaining to our human experience of time.

Such are the main paradoxes which plague our experience of time. They are so intractable that they keep eliciting from each of us the outcry which opens Augustine's meditation on Time: "If nobody asks me, I know; but if I were desirous to explain it to one that should ask me, plainly I know not" (Augustine, *Confessions* XI, XIV).

II. THE NARRATIVE KERNEL

We now turn to narrative discourse and ask how it deals with these paradoxes, whether it provides a solution of its own, and how this solution relates to the lack of speculative resolution which we recognized.

As an immediate rejoinder to our last complaint—we know time when nobody asks us about it, we don't know it when we are asked to explain it—we are first of all overwhelmed by the incredible abundance and variety of narrative expressions (oral, written, graphic, gestural), and of classes of narratives (Myth, folktale, fable, epic, tragedy, drama, novel, movies, comics, as well as history, autobiography, analytical case-histories, testimony of witnesses before the court, and, of course, ordinary conversation). If time-experience is *mute,* narrating is *eloquent.* We find the scattered diversity of story-telling uniquely puzzling. To handle this overwhelming proliferation of narrative forms, the first natural move has been to reduce the number of its classes. A second more artificial move has been to submit them to constructed models. As concerns classification, we may say that the development of our western culture has produced a major dichotomy, that drawn between history and story; i.e. between narratives which claim to be *true,* empirically verifiable or falsifiable, and fictional stories which ignore the burden of corroboration by evidence, and so this dichotomy constitutes an important obstacle for our inquiry, for it makes questionable the claim that these two large classes share some common narrative structures whose temporal features in turn could be easily acknowledged. As concerns the second more artificial device—the recourse to constructed narrative models and codes—it offers a still more dangerous threat to our enterprise, to the extent that the interest in narrative models and codes often results in a trend both in the theory of history and in literary criticism to deny the narrative component, whatever it may be. Some historians speak, after Braudel, of "*histoire non-événementielle*"—eventless history—and many epistemologists in the field of historical knowledge contend that inquiry—in the strong sense of argumentative inquiry—has excluded history writing from traditional narrative forms (folklore, myths, epics and legendary chronicles). On the side of literary-criticism, we have a similar and even more radical trend to submit the intractable variety of narrative forms to constructed models which should be radically a-temporal (e.g. Roland Barthes' "Introduction to the structural analysis of narratives.") Thus, a more manageable deductive approach can be substituted for an impossible inductive approach. The result is that the narrative component as such is identified only with the surface-grammar of the message, with the level of manifestation; whereas only a-chronological codes would rule the level of constitution, to use the vocabulary of A.J. Greimas, the main French exponent of this structuralist treatment.

My suspicion is that both anti-narrativist epistemologists in the field of the theory of history and structuralist literary criticism the theory of story have overlooked the temporal complexity of the narrative matrix in both narrative classes. Because most historians have a poor concept of event—and even of narrative—they consider history as an explanatory endeavour which has severed its ties with story-telling. The underscoring of the surface-grammar in literary narration leads critics to what seems to me to be a false posing of radical choices: either to remain caught in the labyrinthine chronology of the told story, or to radically move to an a-chronological model. This dismissal of narrative as such implies a similar lack of concern in both camps for the properly *temporal* aspects of narrative and therefore for the contribution that the theory of narrative could offer to a phenomenology of time experience. To put it bluntly, this contribution has been almost null because *time* has disappeared from the horizon of the theories of history and narrative. Theoreticians of these two broad fields seem even to be moved by a strange ressentiment against time, the kind of ressentiment that Nietzsche expressed in his *Zarathustra*.

I suggest that we take as the leading thread of our discussion the decisive concept of *plot*. If plot is not the only structure of narrative and if its role in modern narratives has become controversial, at least it may be held that for an inquiry into the temporal aspect of narrative, plot functions as the narrative *matrix*. This emphasis on narrative as plot has several advantages: first, it provides us with a structure which could be common to both historical and fictional narratives. Of course, in order that this advantage remain unchallenged, it must be proved on the one hand that history-writing itself continually arises from this narrative matrix, and therefore that inquiry and explanatory procedures are constantly grafted upon the kind of intelligibility displayed by narratives as plots. In the same way, it must be proved that in fictional narratives, the plot is a basic structure which no a-chronological model is able to generate.

This paper will not give the arguments which support my contention that history writing and fictional narratives not only proceed historically from some common matrix in which plot is an important component, but also continue, in fact, to share in this common narrative structure.

The second advantage which I see in an analysis starting from *plot* is one which would be less dependent on further arguments not developed in this paper: an examination of *plot* may allow us to momentarily bracket one of the main differences between historical and fictional narratives, the difference resulting from the *truth-claim* of history vs. that of fiction. In other words, plot may be seen as pertaining to the *sense* of narrative as distinct from its *reference*. This distinction must not be emphasized too strongly, since the complete meaning of the most fictional narrative cannot be assessed without taking into account its relation to the real world, whether it be a relation of imitation in the narrow sense of copying, or an imitation which encorporates such complexities as irony, decision, conscious distortion and negation, and so on. Ultimately it is our very inquiry into the function of narrative as *shaping our temporal experience which will compel us to go beyond the obvious but provisory opposition between the direct truth-claim of history concerning past events, and the indirect truth-claims of fictional narrative which are implied in the various forms of the mimetic function.*

Such are the two advantages of an inquiry into the narrative structures common to history and fiction starting from plot as the most significant of those structures. I will say nothing of the third and last advantage, that plot displays some remarkable temporal structures which help us to bridge the gap between an inquiry into narrativity and an inquiry into temporality. The description of these temporal structures is in fact the main purpose of this paper, and I freely acknowledge that the desire to disentangle these temporal implications was the main motive which led me to pick out "plot" as my main topic. This implication is already suggested by the fact that the notions of historical *event,* as a temporal concept, and *plot,* as a narrative concept, are mutually definable. To be historical, an event must be more than a singular occurance, a unique happening. It receives its definition from its contribution to the development of a plot. Reciprocally, a plot is a way of connecting event and story. A story is *made out of* events, to the extent that plot makes events *into* a story.

This notion of events made *into* story through the plot immediately suggests that a story is not bound to a merely chronological order of events. All narratives combine in various proportions, two dimensions—one chronological and the other non-chronological. The first may be called the episodic dimension. This dimension characterizes the story as made out of events. The second is the configurational dimension, according to which the plot construes significant wholes out of scattered events. Here I borrow from Louis O. Mink the notion of configurational act, which he interprets as a "grasping together." I understand this act to be the act of the plot, as eliciting a pattern from a succession. I am ready to ascribe to this act the character of *the* judgment, and more precisely of reflective judgment in the Kantian sense of this term. (Dray on judgment). To tell and to follow a story is already to reflect upon events in order to encompass them in successive wholes. Such is the dimension which is completely overlooked in the theory of history by the anti-narrativist writers. They tend to deprive narrative activity of its complexity and above all, of its two-fold characteristic of confronting and combining in various ways both sequence and pattern. But this antithetical dynamic is no less overlooked in the theory of fictional narratives proposed by structuralists. They take it for granted that the surface-grammar of what they call the "plane of manifestation" is merely episodic, and therefore purely chronological. They conclude that the principle of order has to be found at the higher level of a-chronological models or codes. Anti-narrativist writers in the theory of history and structuralist writers in literary criticism share the same prejudice. They do not see that the humblest narrative is always more than a chronological series of events and that in turn the configurational dimension cannot overcome the episodic dimension without suppressing the narrative structure itself.

III. TEMPORAL STRUCTURES OF THE PLOT

We may now try to assess the contribution of the theory of narrative to the "solution" of the paradoxes of time experience.

I shall not attempt to do that in a direct and straightforward way. Rather, I shall use as an intermediate step toward this solution the kind of *parallelism* which can be established between the levels of temporalization that we described in the first section following Augustine and Heidegger, and some corresponding levels which have yet to be acknowledged in the temporal structures of the plot. This parallelism between levels of temporality and levels of narrativity will pave the way for understanding the ways in which narratives, on the one hand, are the modes of discourse appropriate to our experience of time; and time experience, on the other hand, is the ultimate referent of the narrative mode. Instead of claiming to grasp directly the connection between narrative as a language-game and temporality as a form of life, this more analytical and piecemeal approach will help us to construe in a meaningful way the unity of time as narrated and narrative as temporal.

The present paper will remain within the boundaries of an analysis of plot as the structure common to both historical and fictional narrative. We shall not attempt to lift the brackets put on the differences between the truth-claims of these two broad classes of narratives. In other words, we shall speak of plot as the *sense* of narrative isolated from its reference.

In spite of its abstract character—or, maybe, thanks to it—the analysis of the plot is already very rich in temporal implications. We have already anticipated some of them in making the distinction between the chronological and the non-chronological dimensions of plot. But the temporal features of the plot will appear in a more articulate manner if we apply to them the grid of the three-fold structure of time experience as temporality, historicity, and within-timeness.

a) The first function of narratives—whether "true" or fictional—is to establish man at that level of temporalization that Heidegger calls "within-timeness"—the time of everyday life. But in spite of its way of locating events "in time," narrative activity already marks the threshold between the existential traits of within-timeness and the abstract representation of time in a linear way. The art of telling makes a meaningful use of most adverbial expressions which characterize this lower level of temporality: "then," "earlier," "later," "until that . . . ," "now that . . . ." When storytellers start telling, everything is already *extended* in time.

It could be said, then, that narrative activity, taken unreflectively, contributes to the dissimulation of the more authentic levels of temporality. It is more appropriate, however, to say that narrative activity *tells the truth* of within-timeness as a genuine dimension of human Care. All the categories which, according to Heidegger, differentiate within-timeness from the other levels of temporality, make sense at the ordinary level of story-telling. The heroes of the narrative "reckon with" time. They "have" or "don't have" time "to" (do this or that). Their time may be "lost" or "won." Furthermore, narratives show men thrown into circumstances which in turn deliver them over to the change of light and night. It is, accordingly, a time in which "datation" obtains, but according to natural measures, like days and seasons, which have not yet been replaced by the artificial measures of astronomy and physics. Of this time it can be rightly said that "we measure time because we reckon with it" and not the contrary. In the same manner, narrative time is public time, not because it is indifferent to man as acting and suffering. The art of telling keeps the public character from falling back into anonymity. It is the time proper to the "being-together" of heroes, antagonists and helpers, the time of action as interaction—"inter-time," if we dare say, not abstract time. Thus narrative time shares with within-timeness the same public character, insofar as neither have yet been levelled to the abstract representation of time.

Finally, the "now" characteristic of within-timeness, which Heidegger defines as the present which "makes present," is not yet the now of linear time. It is the "from now on . . . " of human decision, and the "now that . . . " of human intervention. This present, as Heidegger says, "temporalizes in union with expectation and retention." It is concern understood as "expectation-that-retains."

To summarize this first point, it belongs to a hermeneutics of story-telling to initiate the return from the *abstract representation* of time as linear to the *existential interpretation* of temporality. Story-telling achieves that in a first way by revealing the existential traits of within-timeness over and against the abstraction of linear time.

(b) But narratives do more than establish man—his actions and passions—"in" time. It brings us back from within-timeness to historicity—from "reckoning with" time to recollecting time. As such, the narrative function provides the *transition* from within-timeness to historicity.

What we said earlier concerning the dialectic between the episodic and configurational dimensions of plot may help us to make this new step; we only have to make the temporal implications of this dialectic explicit. Here we hit upon a temporal constitution which is completely overlooked in the theory of action by anti-narrativist arguments and in literary criticism by structuralist claims. Both take it for granted that narrative as such is merely chronological and that chronology means abstract succession. This is why no other device seems to remain open except a subordination of sequential history to explanatory history, on the one hand, or the reduction of the chronology of the narrative message to the a-temporality of narrative codes. What is overlooked in both camps is the tremendous complexity of narrative time. We could display in the following way the temporal dialectic implied in the basic operation of eliciting a configuration from a succession. Thanks to its episodic dimension, narrative time tends toward the linear representation of time in many ways: first, the "then" and "and then," which provides an answer to the question "what next?" suggests a relation of exteriority between the phases of the action. Besides this, the episodes constitute an open-ended series of events which allow one to add to the "then" an "and then," and an "and so on . . . . " Finally, the episodes follow one another in accordance with the irreversible order of time common to human and physical events.

The configurational dimension, in turn, displays temporal features which may be opposed one by one to those 'features' of episodic time. First, the configurational arrangement makes the succession of events into significant wholes which are the correlate of the act of grouping together. Thanks to this reflective act—in the sense of Kant's *Critique of Judgment*—the whole plot may be translated into one "thought." "Thought," in that narrative context, may assume various meanings. It may characterize, according the Aristotle's *Poetics,* the "theme"—in Greek, the *dianoia*—which accompanies the "fable"—in Greek, the *mythos*—of the tragedy. (It may be noticed that this correlation between "theme" and "plot" is the basis of Northrop Frye's "Archetypcal" criticism). "Thought" may also designate the "point" of the Hebraic maschal or of the Biblical parable (hereto, Jeremias observes that the "point" of the parable is what allows us to translate it into a proverb or an aphorism). The term "thought" may also apply to the "colligatory terms" used in history writing; such terms as the Renaissance, the Industrial Revolution, and so on, which, according to Walsh and Dray, allow us to apprehend a set of historical events under a common denominator (here "colligatory terms" correspond to the kind of explanation that Dray puts under the heading of "explaining what"). In a word, the correlation between thought and plot supersedes the mere "then" and "and then" of mere succession. But it would be a complete mistake to consider such a "thought" as a-chronological. "Fable" and "theme" are as closely *tied* as episode and configuration. The time of fable-and-theme, if we may put that in a one-word expression, is more deeply temporal than the time of merely episodic narratives.

Secondly, the plot's configuration superimposes "the sense of an ending"—to use Kermode's expression—on the open-endedness of mere succession. As soon as a story is well-known—and such is the case with most traditional and popular narratives, as well as with the national chronicles of the founding events of a given community—re-telling takes the place of telling. Then following the story is less important than apprehending the well-known end as implied in the beginning and the well-known episodes as leading to the end. Time, once more, is not abolished by the teleological structure of the judgment which "grasps together" the events under the heading of the end. This strategy of judgments is one of the means through which time experience is brought back from within-timeness to repetition.

Finally, the recollection of the story ruled as a whole by its way of ending constitutes an alternative to the representation of time as flowing from the past forward into the future, according to the well-known metaphor of the "arrow of time." It is as though recollection inverted the so-called "natural" order of time. By reading the end in the beginning and the beginning in the end, we learn also to read time itself backwards, as the recapitulation of the initial conditions of a course of action in its terminal consequences. In that way, a plot establishes human action not only within time, as we said at the beginning of this section, but within memory. Memory, accordingly, *repeats* the course of events according to an order which is the counterpart of time as stretching-along between a beginning and an end.

This third temporal character of plot has brought us as close as possible to Heidegger's notion of *repetition,* which, as we said, is the turning point for his whole analysis of *historicity.* Repetition, for him, means more than a mere reversal of the basic orientation of Care toward the future. It means the retrieval of our ownmost potentialities inherited from our own past in the form of personal fate and collective destiny. The question, then, is whether we may go so far as to say that the function of narratives—or at least of some of them—could be to establish human action at the level of genuine historicity, i.e. of repetition. If such were the case, the temporal structure of narrative would display the same hierarchy as the one established by the phenomenology of time experience.

(c) In order to acknowledge this new temporal structure of some narratives, we have to question some of the initial presuppositions of the previous analysis, and above all, of those which rule the selection of the paradigmatic case of narrative in modern literary criticism. Vladimir Propp, in his *Morphology of Tales,* opened the way, by focusing on a category of tales—Russian tales—which may be characterized as complying with the model of the heroic quest. In those tales, a hero meets a challenge—either mischief or some lack—which he is sent to overcome. Throughout the quest he is confronted with a series of trials which require that he choose to fight rather than to yield or to flee, and which finally end in victory. The paradigmatic story ignores the non-chosen alternatives—yielding and losing. It knows only the chain of episodes which leads the hero from challenge to victory. It is not by chance if, after Propp, this schema offered so little resistance to the attempts by structural analysis to dechronologize Propp's paradigmatic chain. Only the linear succession of episodes had been taken into account. Furthermore, the segmentation of the chain had led to the isolation of temporal segments held as discrete entities that were externally connected. Finally, these segments were treated as contingent variations of a limited number of some abstract narrative components, the famous thirty-one "functions" of Propp's model. The *chronological* dimension was not abolished, but immediately deprived of its temporal constitution as plot. The segmentation and the concatenation of "functions" had paved the way for a reduction of the chronological to the logical. In the new phase of structural analysis, with Greimas and Barthes, the a-temporal formula which generates a chronological display of functions transforms the structure of the tale into a machinery whose task it is to compensate for the initial mischief or lack by a final restoration of the disturbed order. Compared to this logical matrix, the quest itself appears as a mere diachronical residue, as a retardation or suspension in the epiphany of order.

The question is whether it is not rather the initial need to reduce the chronological to the logical—a need arising from the method itself—which rules the strategy of structural analysis in Propp's successive phase: first, the selection of the quest as the paradigmatic case, then the projection of its episodes on a linear time, the segmentation and the external connection of his "function," finally the dissolution of the chronological into the logical.

There is an alternative to de-chronologization. It is repetition. De-chronologization implies the logical abolition of time; repetition, its existential deepening. But to support this view, we have to question the implications and even the choice of the paradigmatic cases of narratives in current literary criticism.

Without putting aside the model of the quest, we may emphasize some of its temporal aspects, aspects which have been ruled out by the method itself. Before projecting the hero forward for the sake of the quest, many tales send the hero or heroine into some dark forest where he or she goes astray or meets some devouring beast (*Little Red Riding Hood*), or where the younger brother or sister has been kidnapped by some threatening birds (The *Swan-Geese* Tale). These initial episodes do more than merely introduce the mischief which is to be suppressed. They bring the hero or heroine *back* into a primordial space and time which is more akin to the realm of dream than to the sphere of action. Thanks to this preliminary disorientation, the linear chain of time is broken and the tale assumes an oneiric dimension which is more or less preserved alongside the heroic dimension of the quest. Two qualities of time are thus intertwined: the circularity of the imaginary travel and the linearity of the quest as such. I agree that the kind of repetition involved in this travel toward the origin is rather primitive, if not even regressive, in the psychoanalytic sense of the word. It has the character of an immersion and confinement in the midst of dark powers. This is why this repetition of the origin has to be superseded by an act of rupture, depicted, for example, in the episode of the woodcutters breaking open of the belly of the wolf with an ax. Nevertheless, the imaginary travel suggests the idea of a meta-temporal mode which is not the a-temporal mode of narrative codes in structural analysis. This timeless—but not a-temporal—dimension duplicates, so to speak, the episodic dimension of the quest and contributes the "fairy" atmosphere of the quest itself.

This first mode of repetition must, in turn, be superseded, to the extent that it constitutes only the reverse side of the time of quest and conquest, brought forward by the call for victory. Finally, the time of quest prevails over that of the imaginary travel through the break thanks to which the world of action emerges from the land of dreams—as though the function of the tale was to elicit the progressive time of the quest out of the regressive time of imaginary travel.

Repetition tends to become the main issue of the narrative in the kind of narratives in which the quest itself duplicates a *travel* in space which assumes the shape of a return to the origin. Odysseus' travels are the paradigm of the narrative as travel and return. As professor Eliade writes in *L'Epreuve du Labyrinthe (The Trial of the Labyrinth)* (P. 109) "Ulysses is for me the prototype of man, not only modern man, but the man of the future as well, because he represents the type of the "trapped" voyager. His voyage was a voyage toward the center, towards Ithica, which is to say, towards himself. He was a fine navigator, but destiny—spoken here in terms of trials of initiation which he had to overcome—forced him to postpone indefinitely his return to hearth and home. I think that the myth of Ulysses is very important for us. We will all be a little like Ulysses, for in searching, in hoping to arrive, and finally, without a doubt, in finding once again the homeland, the hearth, we re-discover ourselves. But, as in the Labyrinth, in every questionable turn, one risks "losing oneself" (*se perdre*). If one succeeds in getting out of the Labyrinth, in finding again one's home, then one becomes a new being." The retardation of which Eliade speaks here is no longer the mere "suspension" in the epiphany of order. Retardation now means growth.

The Odyssy, accordingly, could be seen as the form of transition from one level of repetition to another, from a mere fantasy repetition which is still the reverse side of the quest, to a kind of repetition which would generate the quest itself. With the *Odyssy* the character of repetition is still imprinted in time by the circular shape of the travel in space. The temporal return of Odysseus to himself is supported by the geographical return to Odysseus' birth-place, Ithica.

We come closer to the kind of repetition suggested by Heidegger's analysis of historicity with stories in which the return to the origin is not only a preparatory phase of the tale and is no longer mediated by the shape of the travel back to Ithica. In these stories, repetition is constitutive of the temporal form itself. The paradigmatic case of such stories is Augustine's *Confessions*. Here the form of the travel is interiorized to such a degree that there is no longer any priviledged place in space to return to. It's a travel "from the exterior to the interior, and from the interior to the superior." (*"Ab exterioribus ad interiora, ab interioribus ad superiora."*) The model created by Augustine is so powerful and enduring that it has generated a whole set of narrative forms down to Rousseau's *Confessions* and to Proust's *Le Temps Retrouvé*. If Augustine's *Confessions* tell "how I became a Christian," Proust's *Le Temps* tells "how Marcel became an artist." The quest has been absorbed into the movement by which the hero—if we may still call him by that name—becomes *who he is*. Memory, then, is no longer the narrative of external adventures, stretching along episodic time. It is itself the spiral movement which, through anecdotes and episodes, brings us back to the almost motionless constellation of potentialities which the narrative retrieves. The end of the story is what equates the present with the past, the actual with the potential. The hero *is* who he *was*. This highest form of narrative repetition is the equivalent of what Heidegger called Fate—individual Fate—or destiny—communal destiny—i.e. the complete retrieval in resoluteness of the inherited potentialities in which Dasein is thrown by birth.

At this point the objection could be made that only fictional narrative, and not history, reaches this deep level of repetition. I do not think this is the case. It is not possible to ascribe only to *inquiry*—as opposed to traditional narrative—all the achievements of history in the overcoming of legendary accounts, i.e. the release from mere apologetic tasks related to the heroic figures of the past, the attempt to proceed from mere narrative to truly explanatory history, and finally the grasp of whole periods under a leading Idea. We may wonder whether the shift described by Mandelbaum in the *Anatomy of Historical Knowledge* from sequential history to *explanatory* history does not find its complete meaning in the further shift from explanatory to what he calls *interpretive* history.

> While an interpretive account is not usually confined to a single cross-section of time but spans a period . . . the emphasis in such words is on the manner in which aspects of society or of the culture of the period, or both, fit together in a pattern, defining a form of life different from that which one finds at other times or in other places" (pp. 39f.).

Am I stretching the notion of *interpretation* too far, if I put it in the Heideggerian terms of repetition? Professor Mandelbaum may dislike this unexpected proximity to Heideggerian ideas. I find, nevertheless, some confirmation and some encouragement to take this daring step in the profound analysis of action that Hannah Arendt gives in her brilliant work *The Human Condition*. As is well known, Arendt distinguishes between labor, work and action. Labor, she says, aims merely at survival in the fight between man and nature. Work aims at leaving a mark on the course of things. Action deserves its name when, beyond the concern for submitting nature to man or for leaving behind some monuments witnessing to our activity, it aims only at being recollected in stories whose function it is to provide an identity to the Doer, an identity which is merely a *narrative identity.* Thus history *repeats action* in the figure of the memorable.

Such is the way in which history itself—and not only fiction—provides an approximation of what a phenomenology of time experience may call repetition.

* * * * *

My conclusion will be less a summary of the previous analysis than a set of suggestions for further inquiry.

First, to what extent can we say that narrative activity solves the ambiguities and paradoxes of our experience of time? It can achieve no speculative resolution, but perhaps may achieve a *poetic* one. By that I mean that by telling stories and writing history we provide "shape" to what remains chaotic, obscure and mute. But the complete account of this function of narrative as *shaping* time would imply a development and a reformulation of all our previous analysis in terms of *productive imagination,* to use Kant's terminology. We should have to show that historical narrative and fictional narrative *jointly* provide not only "models of" but "models for" articulating in a symbolic way our ordinary experience of time. This has not been done in this paper.

And this could not have been done, because we have not overcome the abstraction thanks to which we were able to isolate the *plot* as the structure of *sense* common to history and to fiction. In order to show the way in which history and fiction *jointly* shape our experience of time, we should have to proceed from *sense* to *reference*, i.e. to show how the obvious differences between history as "true" story and *fictional* story work together, so to speak, beyond the asymmetry in truth-claims. This task cannot be achieved within the boundaries of a mere epistemology of narrative forms. It requires a hermeneutical approach able to encompass both forms under a broader concept of truth than the epistemological one which is assumed when we oppose *"true" story to fictional story. I don't deny that this task can be partially achieved within a framework which can still be called epistemological. On the one hand, it can be shown, with such writers as Hayden White, that history-writing is more fictional* than positivist writers would allow us to say. On the other hand, it can be shown, with a whole school of literary criticism arising from Aristotle's *Poetics,* that fiction is more *mimetic* than the same positivistic trend of thought would acknowledge. Between a fictional history and a mimetic poetry some striking convergences could therefore be discerned, and an intersection between the direct referential claim of "true" story and the indirect referential claim of fictional story could appear as a plausible horizon for inquiry into narratives at large. But I have no difficulty in acknowledging that this *intersection* remains only a plausible horizon for research as long as we keep talking in epistemological terms. The gap between historical and fictional narrative could only be bridged if we could show that both are grounded in the same basic temporality which provides to repetition itself an existential foundation. Then, it could be shown that the reasons for which we write history and the reasons for which we tell stories are rooted in the same temporal structure that connects our "elan" towards the future, our attention to the present and our capacity to emphasize and to recollect the past. Then "repetition" would no longer appear as a dubious procedure divided between fictional repetition and historical repetition. It would be able to encompass history and fiction to the extent that both are rooted in the primordial unity between future, past and present.

But this ontological part of the inquiry is, unfortunately, beyond the scope of this study.

The Common Presuppositions of Hermeneutics and Ethics: Types of Rationality Beyond Science and Technology

KARL-OTTO APEL
University of Frankfurt

I. EXPOSITION OF THE PROBLEM: THE WESTERN COMPLEMENTARITY-SYSTEM OF VALUE-FREE RATIONALITY AND PRE-RATIONAL VALUE-DECISION AS A CHALLENGE TO REASON

The title and the sub-title of my paper are meant to point to a direction of philosophical reflection that, I shall submit, must provide the key for a foundation of ethics as well as a foundation of hermeneutics as the methodology of "understanding" in the humanities or *Geisteswissenschaften.* The question, I submit, is whether there are types of rationality other than those of value-free explanatory science and its transposition into technology, including social engineering.

A famous answer was given to this question by the great sociologist Max Weber at the beginning of this century: His answer was ambiguous in a sense with regard to my question. For, on the one hand, he propagated a type of sociology that would be founded on "understanding"—understanding, that is, of individual persons, their actions and works as well as institutions, different cultures, and historical epochs, and even trends of progress in history in the light of so-called "relations to values" ("Wertbeziehungen"). Thus far Max

Weber followed the German philosophers of the "*Geisteswissenschaften*" or, respectively, "*Kulturwissenschaften*" who made a methodologically relevant *distinction* between causal and statistical "explanations" of the phenomena of nature in the light of general laws, on the one hand, and "understanding" the individual phenomena of human history or culture in the light of special complexions of general values, on the other.[1]

Max Weber, however, also proposed a strictly negative answer with regard to my question as to possible types of rationality beyond value-free science and technology. And this answer has become paradigmatic, I suggest, for the present system of Western idology. Weber restricted the scope of methodical understanding to "value-free" understanding which he centered around the "ideal type" of "purposive-rational understanding" of "purposive-rational actions".[2] Now "purposive-rational actions" may also be called "instrumental actions;" and in those cases where these actions are successful, they may be analysed or reconstructed as being based on successful transpositions of the if-then-rules of nomological science into the if-then-rules of technological prescriptions. Hence Max Weber thus restricted the business of methodical understanding to the attempt of grasping the (value-free) *technological means-ends-rationality* beyond the human actions. And it is this idea of *instrumental rationality* which indeed constituted Weber's paradigm of rationality in a restrictive sense.

It has to be pointed out, though, that for a *purposive-rational understanding* in sociology it is not necessary to fulfill the maximal requirement of making sure that the agent succeeded in transposing nomological rules into his technological maxims about means-ends-relations. In order to *understand* his actions in the light of that type of instrumental rationality, it is enough to make sure that it was rational for the agent to act as he did under the presupposition of *his* aims and *his* beliefs about means or ways or strategies as being suited to reaching his aims. Thus it becomes the empiric-hermeneutic business of understanding to hypothetically find out and verify those goal-intentions and means-beliefs on the side of the agent, in the light of which his actions can be understood as being rational in the sense of technological means-ends rationality.

This requirement of finding out the aims and beliefs of the agent also makes it understandable that Max Weber's restriction of the idea of rationality to technological means-ends-rationality did not prevent him from setting up the task of understanding the particular aims and beliefs of the individual agents in the light of "value-relations." For his opinion was not that the sociologist should evaluate the last aims or beliefs of human agents in the light of, say, a binding order of values or a correct world-view. Rather the historical complexions of general values and world-views were only considered as indispensable heuristical horizons for finding out the aims and beliefs of the agents under the presupposition of which the sociologist could understand the actions themselves as being intelligible in the light of means-ends-rationality. Hence, on Weber's account, the formal rationale of technological means-ends-rationality was the only normative standard in the light of which a sociologist had to evaluate human actions, in order to understand them in a value-free way.

This issue of Weber's methodology of "understanding" was in perfect accordance with his (more or less implicit) philosophy of history. For in the context of his own reconstruction of the history of Western civilization, he started out from the heuristic hypothesis that at least this part of history could be conceived of as a continuous progress of "rationalization" and at the same time, as a process of disillusionment or, as he liked to say, "disenchantment" ("*Entzauberung*"). By "rationalization" he understood the progress in putting into force means-end-rationality in all sectors of the socio-cultural system, especially in the sphere of economics and bureaucratic administration, under the constant influence of the progress in science and technology. By the process of "disillusionment" or "disenchantment," on the other hand, Weber understood, among other things, the dissolution of a commonly accepted religious or philosophical value-order or world-view. And he was prepared to draw practical consequences from this development for his personal world-view in so far as he suggested that a rigorous and sincere thinker had to accept the following insight: Human progress in the sense of "rationalization" has its complement in giving up the idea of a rational assessment of last values or norms in favor of taking recourse to ultimate pre-rational decisions of conscience in face of a pluralism, or, as Weber said, "polytheism" of last norms or values.[3]

By this last philosophical conclusion, Max Weber became one of the first exponents of what I would call the present *complementarity-system of Western (liberal) ideology*. According to this system, one has to make a clear-cut distinction between two spheres of life: In the *public sphere* only the laws and rules of scientific-technological rationality must be acknowledged as intersubjectively valid and hence as binding, whereas all legal and moral norms must be traced back to mere conventions.

Hence in the public sphere of life scientistic positivism supplemented by a certain type of instrumentalist pragmatism becomes, so to speak the leading philosophy. In the *private sphere*, on the other hand, the decisions behind the conventions of the public sphere may be further traced back to pre-rational personal decisions, possibly even to the decision of the majority of all citizens that value-free scientific-technological rationality must be acknowledged as binding in the public sphere. Hence in the private sphere of life a certain type of (religious or atheistic) Existentialism must be the leading philosophy. And pragmaticist Positivism, on the one hand, and irrationalist Existentialism, on the other hand, by no means contradict each other within the frame of this ideological system, but, on the contrary, they complement each other in the way of a certain division of labor with regard to the public and the private sphere of life, respectively.

This *complementarity-system of value-free rationality and pre-rational value-decisions*[4] may easily be recognized as corresponding to the earlier liberalist separation between the state and the church according to which the state has to be neutral with regard to the question of last norms, values or philosophical truths. And at first sight it could really appear as though the complementarity-system thus far outlined could figure as the necessary and sufficient alternative of the so-called free West in comparison with the Eastern *integration-system* where the state, or the party, has the function of putting into force the dogmatic truth of a non-value-free superscience, called "Diamat," as the binding norm for the spheres of both public and private life.

I do not think that the outlined ideological system is in fact necessary and sufficient as an alternative to dogmatic totalitarianism. On the contrary, I consider it to be an extreme challenge to human reason with regard to the question as to types of rationality beyond science and technological means-ends-rationality. In the present paper, I will try to take up that challenge with regard to both of Max Weber's tenets: viz. with regard to the claim that hermeneutic understanding must be "value-neutral" by taking the formal standard of means-ends-rationality as criterion of rational understanding (or "rational explanation," as it is called even nowadays);[5] and with regard to the claim that ultimate rational foundation of ethical norms has to be replaced by recourse to ultimate pre-rational private decisions. I would like to suggest from the outset that there is a common, rational basis for hermeneutic and ethic rationality, i.e. for the methodological rationale of a non-value-neutral hermeneutic understanding of human actions in history, and of history as a whole, and, on the other hand, of grounding ethical norms by means of "*Versténdigung,*" i.e. by means of communicative understanding concerning the meaning of validity-claims and by coming to consensus about intersubjectively binding norms.

However, since, as I know, there are very strong pre-conceptions predominant in sincere liberal thinkers toward believing that my talk about an ultimate foundation of norms and non-value-free interpretation of history as a whole must amount to precisely that type of historicist super-science that is the basis of Eastern totalitarianism, I will first try to undermine the suggestive plausibility of the outlined Western complementarity system of value-neutral rationality and pre-rational value-decisions by exposing the weak points of the system.[6]

1. Firstly, with regard to the public sphere of life, the following aporia of pragmaticist positivism becomes manifest, I suggest: It is simply not true that a thorough value-pluralism or, for that matter, skepticism can be the basis of public decision-procedures in a liberal-democratic state, such that the state itself could and should be completely neutral with respect to the value-decisions behind the conventions about the constitution and about all sorts of positive laws. It is true, of course, that, in a democratic state, all kinds of public rights and duties, especially those of the state-organs, must be grounded in the last resort by conventions of the major citizens. But in order to make such an order of the public life possible, a fundamental consensus about certain normative conditions of the possibility of value conventions has to be *presupposed,* and beyond that of course one even more fundamental norm: the norm that all positive legal norms or laws should be grounded by valid conventions. These points, I suggest, indicate the moral conditions of democratic pluralism.

2. Secondly, with regard to the private sphere of life, one may point to the following dilemma of irrationalist Existentialism. At the moment where the individual persons, following M. Weber (and, for that matter, also K. Popper)[7] might consider all their ultimate norms or maxims of acting and evaluating to be based on pre-rational decisions, or "acts of faith," with no rational claim to intersubjective validity—at that moment the idea of conscience and hence of decisions of conscience, which makes up the emphatic point of secularized Christianity behind modern Liberalism, is bound to dissolve itself. For at this point, no coercive motive is left over for grounding one's life-decisions on personal maxims at all, e.g. for grounding one's political behaviour on the will to striving for conventions with other people and, especially, on the will to keeping the conventions, i.e. to one's promises to the other persons, even in cases

where one's personal interests suggest breaking the conventions since no sanctions have to be feared.

It is not true that, under the presupposition that my ultimate norms or maxims are based on pre-rational private decisions, there could still be a reasonable motivation for my decisions to be derived from my reflections about the practical consequences of my decisions. This standard argument of Popperians rests on the same fallacy as their proposal to replace the allegedly impossible ultimate foundation of ethics by a method of critical comparing of different ethical theories in the light of their probable consequences.[8] For this good idea presupposes of course that the person in charge of comparing is in possession of an ultimate normative standard. (The same fallacy, by the way, is lying at the ground of that dissolution of Popperian "Critical Rationalism" that manifests itself in the later work of P. Feyerabend;[9] for as far as I can see, the pitiful logic of Feyerabend's development from setting up the fruitful principle of proliferation of theories up to the last finding that "any-thing goes" is simply determined by the fact that he extrapolates the proliferation principle again and again, but, from the outset, does not care for the normative standards of comparing theories).

In the face of this situation, one may set up the thesis that at the moment where persons would base their ultimate valuations on nothing but pre-rational decisions, the internal motivation of people's decisions by ethical maxims must break down and give way to external conditioning, say by political or economic propaganda or other quasi-causal motives.[10]

These few remarks may suffice in order to show that we have good reasons for doubting the tenability of the outlined complementarity-system of Western ideology. Hence let us turn back to Max Weber as one of its fathers and try to examine more closely his general claim that there is no rational basis either for evaluation in the "human sciences" or for grounding ultimate ethical norms.

II. MAX WEBER AND THE NON-VALUE-FREE PRESUPPOSITIONS OF HERMENEUTIC RATIONALITY

To begin with Weber's demand for *value-neutrality* of sociological understanding and explanation, there are no doubt some meanings of this principle that commend it as valid for all social scientists:

1. It must of course be granted that valuation in the sense of wishful thinking should not be permitted to conceal from a sociologist impalatable facts.

2. Rather less trivial is the following point that also must be granted: No sociologist can legitimately derive intersubjectively valid norms—and thereby a rational basis for valuating—from facts, i.e. from those norms that may be understood as being factually valid in certain societies.

The fact that the books of sociologists and even more the books of historians, being written in ordinary language, are usually full of more or less implicit valuations,—this undeniable fact has been plausibly explained by Max Weber in the following way: Such writers, he suggests, are implicitly taking recourse to general norms or values that are still tacitly acknowledged as belonging to the Western cultural tradition or at least as part of special national traditions. But Weber hurries to add that this recourse to traditional values or norms is not legitimate for a social scientist; and the reasons he provides for this verdict are indeed crucial for his philosophical position:

First Weber demands that a scientist should lay open to himself and to the public the last presuppositions of all public norms and the factual evaluations, and he emphasizes that this exploration of the value-systems behind factual valuations can itself be accomplished by value-free procedures of understanding and thus is part of scientific enlightenment. With this point of Max Weber's I fully agree.

But then Max Weber goes on to maintain that by tracing back all factual valuations to their tacitly acknowledged presuppositions, the modern sociologist-historian must come to realize that there is no common rational basis for norms or evaluations but only a "polytheism" of last values the claims of which are even fighting against each other in a similar way as the ancient Gods used to do. At this point Max Weber's meta-theory of value-free sociology seems to converge in the last resort with that historism-relativism that was already tendentially prefigured by L. Ranke's famous dictum that "all epochs of history stand in an immediate relation to God," and was finally displayed by W. Dilthey, especially by his projected transformation of philosophy into an empiric-hermeneutic theory of world-views ("Weltanschauungslehre"). And it may be added that at present a similar relativism has been become very influential as a consequence of the later Wittgenstein's suggestion of incommensurable paradigms of language-games and corresponding "forms of life." Even the idea of progress in the history of natural science has been questioned in the light of this new paradigm-relativism, as is well known.[11]

Now let us take up the challenge at this point. And let us pick up the cue (catch-word) that is offered by Max Weber's demand that we trace our evaluations back to their last presuppositions. More precisely: let us e.g. single out Max Weber's own demand for value-neutrality as a normative judgment and inquire into ultimate presuppositions that might render this judgment meaningful. The answer to this question, as far as I can see, must be something like this: If, and only if, scientific statements are value-neutral, can they claim objectivity, i.e. intersubjective validity; and only in this case is scientific progress towards reaching the truth possible. Hence it is for the purpose of securing the conditions of the possibility of that "rationalization-process" that is made possible by scientific progress towards the truth, that value-neutrality has to be demanded by M. Weber.

What is interesting at this stage of our reflection is the fact that M. Weber can only meaningfully postulate value-neutrality of scientific statements in the name of a non-value-neutral principle of his own, which he cannot relativize thus far but has to presuppose as a last regulative principle for interpreting the course of history as a whole. If this is true, then the claim that hermeneutic interpretation of history as a whole and historical actions as part of history should be value-neutral becomes already paradoxical. For Max Weber's own principle of interpreting history as a "rationalization" process based on scientific progress toward the truth provides the very paradigm for a non-value-neutral type of hermeneutic interpretation.

It may be said that this type of hermeneutic interpretation is only a pre-scientific one, that is required for the selection of interesting topics of science but should give way to value-free procedures as soon as scientific cognition begins. But this scientistic standard argument seems to me to be completely mistaken.

To put it more clearly: I would agree that descriptions and nomological explanations in natural science (more precisely, in Physics) must be value-neutral and hence must be sharply separated from pre-scientific evaluations that usually are involved in the interests of scientists and the society that are considered as external steerings of scientific enterprises. The same is true with respect to the descriptions and quasi-nomological explanations of those un-historical quasi-naturalist types of social science that serve as basis for social-engineering.[12] The reason for the necessary link between value-neutrality and objective validity in these cases is obviously provided by our interest in objectifying and controlling nature and human quasi-nature in the service of technology. For, as an internal cognitive interest that constitutes the leading questions in physics and in quasi-naturalist social science, our very interest in controlling compells us to refrain from evaluating the facts that are to be controlled. They have to be analyzed not in the light of normative principles which cannot be falsified, but in the light of empirical law-hypotheses which can be falsified.

But this consideration cannot hold good for a hermeneutic reconstruction of our own innovative teleological actions that make up at least that part of history that may be interpreted as a "rationalization-process" and hence as progress in the sense of Max Weber. With regard to this enterprise, it is not meaningful to set up a methodological separation between the basic value-perspective of interpreting history as a whole and the critical interpretation of all single innovative—or, for that matter, regressive—actions in the light of the basic value-perspective.

The correctness of this conclusion is reconfirmed, I would suggest, at least implicitly by Karl Popper. For he has clearly recognized that historically relevant innovative actions, as e.g. creative achievements of scientists, cannot, in principle, be the object of predictively relevant nomological explanations, but must be the object of "ex post understanding".[13] (Thereby Popper has in fact acknowledged the well interpreted explanation-understanding dichotomy of Dilthey and the Neo-kantians, i.e. Windelband and Rickert. This has to be emphasized against some of Popper's disciples who are still deeply biased by scientistic reductionism). Now it is clear that "ex post understanding" of actions as being innovative or creative in Popper's sense, cannot be conceived of as value-neutral.

This fact is illustrated very nicely by the recent controversy between Th. Kuhn and the Popperians with regard to the history of science. Take e.g. Th. Kuhn's initial self-understanding as an empirical social scientist who wished to describe and explain just the factual actions of scientists in the light of their factual causally effective motives, as e.g. the "social-psychological imperatives" of the different groups or communities.[14] If this understanding of the history of science were correct, then, of course, no challenge of the history of science to the philosophy of science could be imagined. For a normative methodology cannot, in principle, be called into question by purely contingent facts concerning the behavior of scientists. But, after a closer look at the controversy, one can easily see that even Th. Kuhn is not primarily interested in a value-neutral description and causal explanation of just all the facts, but rather in a hermeneutical reconstruction of those *reasons of successful moves* in the history of science that seem not to be compatible with the norms of the methodologists thus far proposed. Even the famous distinction between "normal science" and "revolutionary science" may be evaluated in the light of such a normatively relevant attempt of understanding the reasons for a successful, i.e. progressive, development of science.

I do not wish, however, to defend the details of Kuhn's conception of scientific progress which, on the contrary, I would contest. I am only emphasizing that history of science as a relevant hermeneutic discipline cannot be reduced to a value-neutral type of quasi-naturalist social science. This thesis is in fact much better illustrated by I. Lakatos' conception of a reconstructive history of science which corrects Kuhn's quasi-naturalistic self-understanding by introducing the important distinction between reconstructing the internal history of science, i.e. of scientific progress towards the truth, and explaining the contingent facts by external motivations in those cases where they can no longer be understood in the light of good scientific reasons.[15]

In any case, the history of science (which we may note parenthetically represents in our day the best available example for a "hermeneutic" "*Geisteswissenschaft*" since its business is not to psychologically explain understanding but to hermeneutically understand explanations)[16] cannot be understood with regard to its cognitive accomplishments under the Weberian pre-conception of value-free understanding. On the contrary, it presupposes as a condition of its possibility at least the fundamental norm of striving for the truth and, on the basis of this norm, a thoroughgoing value-perspective for interpreting history as a whole and innovative or regressive actions of man within history.

At this point, however, one could try to defend Max Weber's position by two objections which seem to be interconnected in the last analysis:

1. On the one hand, one could point out that the universal value-perspective of scientific progress toward the truth does not provide a basis for ethical valuation, since it is in fact only a value-principle that is immanent to science.

2. On the other hand, one could claim that for the same reason, i.e. because of its being immanent in science, the principle of truth-progress cannot figure as an ultimate normative principle; for whether science ought to be or not seems to be a matter of our free will, hence of a last pre-rational decision.

These last arguments, which mark a common background-position of Max Weber and Karl Popper and, of course, a cornerstone of the complementarity-system of Western ideology, lead us to the dramatic culmination-point of our inquiry, so to speak. For, in order to show the common presuppositions of hermeneutic and ethic rationality, it is indeed necessary, I think, to overcome what appears at first sight to be the plausibility of the two last mentioned arguments.

In what follows I shall therefore try to show first that the pursuit of truth is by no means only immanent to the special enterprise of science, although it indeed provides a normative legitimation for science. The pursuit of truth, I would claim, is rather an implication of one of the validity-claims that are in fact immanent in all human communication and hence in the hermeneutic rationality of communicative understanding which makes up a *presupposition* of scientific rationality.

Secondly, I will try to show that the a priori of communication which cannot be called into question, since being the apriori of argumentation it is always presupposed by questioning, provides the normative foundation not only for the pursuit of truth but also for ethics.

III. THE APRIORI OF COMMUNICATION AS PRESUPPOSITION OF SCIENCE AND AS ULTIMATE FOUNDATION FOR HERMENEUTIC AND ETHICAL RATIONALITY

Let me start out from the following interesting fact: K. Popper, as before him C. S. Peirce, has clearly recognized that the scientific pursuit of truth presupposes the validity of an *ethics.*[17] Now this insight does not help us very much as long as the acknowledgment of such an ethics, together with the option for the pursuit of scientific truth, must be traced back to a *last pre-rational decision.* In order to avoid that aporia, I must first introduce the hermeneutic rationality of communicative understanding.

Let us, for this purpose, ask the question: Why is it the case that science must presuppose an ethics? A first answer to this question might be: science presupposes an ethics because it is a kind of *purposive-rational activity.* I would not deny that purposive-rational activities presuppose an ethics in order to ground their purposes; but this consideration does not help us to find a foundation for ethics, for it only reproduces the problem of grounding the decision that science ought to be. In fact the circumstance that science is a purpose-rational activity does not show us *why* it must presuppose certain ethical norms and among them the duty of the pursuit of truth. The reason for this failure may be exposed by the following consideration:

If it were true that science be only a matter of the *subject-object-*

relation, as it has been suggested by modern epistemology, so that in principle *one man alone* could perform science, then one could not understand that science should presuppose certain ethical norms. For in this case it could appear to be only a matter of perceptual evidence just for my consciousness and of theory-formation by logical operations—say of deductive, inductive and abductive inference—which could be traced back in principle to the competence of a single subject of cognition. In this case, science would not presuppose the ratio of a certain ethics because it would not presuppose the *hermeneutic rationality of communicative understanding between co-subjects of cognition.*

But the "methodical solipsism" (Husserl) of modern epistemology was overthrown not only by the later Wittgenstein's argument against the possibility of a "private language," but earlier and even more clearly by C.S. Peirce and the almost forgotten American social philosopher J. Royce.

Royce in his last work,[18] elaborating on Peirce's semiotics showed that all cognitive operations upon nature, being bound up with the use of publicly understandable signs, must constantly presuppose the communicative processes of fixing the nominal value of our concepts by sign-interpretation. And he made clear, that this communicative process of sign-interpretation has the same structure (of mediating the three dimensions of time) in the case of solitary thinking as well as in the case of hermeneutic interpretation of the cultural tradition. In both cases the triadic structure may be represented by the scheme: A interpretes to B the meaning-intention of C. By this insight of Royce's it has been shown, I suggest, that the hermeneutic rationality of communicative understanding, which lies at the heart of the philological-historical disciplines, does not compete with the scientific rationality of explanatory arguments but is rather to be conceived of as *complementary* to it.[19]

I can only suggest at this place that the whole controversy about the relationship between *explanation* and *understanding* might be lifted to a more appropriate level of reflection if one substitutes this view of *complementarity* for the usual question as to whether the logical structure of rational understanding, e.g. in the sense of M. Weber or W. Dray or G.H. von Wright, might be *reduced* to the structure of causal explanation, say e.g. in the sense of C.G. Hempel.[20]

For if our question is why a certain action considered as an *event* in space and time came about so that it might have been expected, then it is a question for a *theoretical explanation* in Hempel's sense; and in this case, where the theoretical subject-object-relation of objectifying science is presupposed, one can hardly dispute the possibility of a formal reduction to the DN or IN-Model of Hempel, notwithstanding the difficulties with providing genuine law-hypotheses.

If, however, our question concerns the *reasons* (i.e. maxims, goal-intentions and beliefs) in the light of which the actions of other people can be *understood,* i.e. interpreted as entities of the socio-cultural sphere of reality; especially, if we wish ex post to understand in this way the innovative actions or works of the classics of culture, then we are taking our departure from another *cognitive interest* which is complementary to that of theoretical explanation because our communicative understanding of other people is always presupposed by scientific cognition as Royce, among others, has shown. Hence the hermeneutic rationality of the *"Geisteswissenschaften"* as an autonomous type of methodological rationality can only be clearly understood if it be traced back to the *subject-co-subject-relation* of communicative understanding that is complementary to the *subject-object-relation* of theoretical explanatory science.

But what is the significance of this insight into the nature of hermeneutic rationality for our question why science must presuppose an ethics? The provisional answer to this question at this place might be:

Science must presuppose an ethics because science must presuppose communicative understanding between persons as co-subjects of cognition and because communicative understanding between persons presupposes an ethics. In other words: Science must presuppose an ethics because cognition cannot be understood on the basis of "methodical solipsism" as was attempted from Descartes through J. Locke to the *Cartesian Meditations* of E. Husserl. Science must presuppose ethics because truth is not only a matter of *evidence* for my consciousness, but more over a matter of *intersubjective validity* to be testified to by a *grounded consensus* about the coherence of evidences in the community of investigators.[21] Hence science must presuppose communicative understanding between persons as co-subjects of agreements about truth; and communicative understanding between persons presupposes certain ethical norms.

But at this stage of our argument, the second objection—sc. the Weber/Popper-objection against *ultimate foundation*—could again be brought forward. For one may still insist that the validity of those ethical norms that are presupposed within the community of investigators depends on my pre-rational decision, i.e. on my opting for the pursuit of truth by science.

In order to reply to this objection, one may first point to a preliminary result of our argument that must hold in any case. Max Weber's suggestion that the growth of scientific-technological rationality, based on value-neutral objectivity, must tendentially *exclude* the intersubjective validity of an ethics and hence must *preclude* all claims of a rationality beyond science and technology, must be false. For it has been shown already that science indeed does not exclude but presuppose the intersubjective validity of some type of ethics. This insight seems to me already to mark an important break with respect to the Western complementarity system of value-free rationality and pre-rational evaluation. But this break may be widened into a complete conquest and dissolution (suspension) of the system by the following consideration:

Although communicative understanding in the broad sense of striving for agreement about meaning and truth[22] is in fact an indispensable pre-condition of science, it is by no means a pre-condition only for science. It must rather be considered to be a pre-condition of communicative interaction between human beings and especially as a necessary pre-condition of argumentative discourse. Now, without communicative interaction human life must break down. And if some tough-minded philosopher should insist that he, after all, could make a pre-rational decision in favour of the break-down of life, one may grant that but add that he nonetheless cannot *consistently argue* against the intersubjective validity of some foundational norms of ethics. For, as long as he *argues*—for whatever position—so long must he presuppose the ethics of the ideal communication-community which is always anticipated—more or less counterfactually—by the speech-acts of meaningful arguing.[23] Even the solitary decision of the man who refuses to argue and thereby rejects all rational norms can only be *understood as a meaningful act of decision* in face of an alternative so long as one presupposes that frame of hermeneutic and ethic rationality which he wishes to deny.

Thus the following seems to become clear. The act of decision or faith that is claimed as ultimate basis for ethics by M. Weber and K. Popper—together, by the way, with the existentialist secularization of the Franciscan theology of the pre-rational will of God!—,—this act of decision may at best be called pre-rational in so far as it cannot be *enforced* by rational arguments. This is undeniable with respect to any act of decision, but it is quite another thing than proving that the decision cannot be grounded by rational arguments. For even by valid rational arguments we can never enforce any practical decision, say that of following certain norms. Otherwise grounding ethical norms would prevent freedom and responsibility and hence would destroy the very meaning of ethics. This shows that we must not confuse the freedom of practical decisions with the thesis of *decisionism*, viz. that an ultimate foundation of ethical norms is impossible and hence an ultimate decision has to step in the place of an ultimate foundation.

Against this confusion, I wish to propose the following thesis: An ultimate foundation of ethical norms can be provided by *transcendental-pragmatic reflection on the normative pre-conditions of meaningful arguing.* That is a quasi-Cartesian argument that differs from Descartes' argument by its overcoming the "methodical solipsism" of the "ego cogito, ergo oum" in favor of the apriori of argumentative discourse, i.e. of communication and the communication-community. The so called *pre-rational decision,* on the other hand, be it affirmative or be it negative with respect to the norms of ethical reason, does not affect the validity of ethical norms but is only—indeed—presupposed for the *practical realization* of ethical reason. This, of course, remains a matter of the *good will,* even if an ultimate rational foundation of ethics cannot be disputed without actual self-contradiction.

But it might still be asked: what should be the meaning-content of those ethical norms that are presupposed by any argument, and, in what respect should it make up a common presupposition of ethics and hermeneutics. In order to answer this question, let us first try to clarify the relationship between argumentation and communication with the aid of modern speech-act-theory.

There is a conception of *argumentation* that equates it with the logical operations by which we derive propositions from propositions or, at best, with those synthetic inferences like abduction and induction by which we set up and/or confirm hypotheses (i.e. synthetic propositions) by drawing cognitive information from sense-data which have the semiotic status of indices. Now after what has been said about the methodical solipsism of traditional epistemology, it seems clear that on the basis of this conception of argumentation it is not possible to show that argumentation implies communication and hence ethics. Rather it would appear that communication is an empiric pragmatic function of language from which we might completely abstract if we wish to understand *argumentation.* As far as argumentation is concerned, one might say, only the representation-function of propositions as possible "truthbearers" is relevant, whereas the communicative function of language, which we have in common with the animals according to Karl Bühler,[24] may be thematized by empirical disciplines like socio-linguistics.

Now this conception of languge and argumentation, I would argue, is an obsolete (or superseded) paradigm. Drawing consequences from the work of Austin and Searle, one should say that it is the *performative-propositional double-structure* of sentences which express whole speech-acts, as e.g. questions and assertions, that constitutes the structure of argumentation and hence the priority of human language rather than merely the representation-function of propositions. For, as Habermas has shown,[25] it is only by recourse to the human *truth-claim,* as it is made explicit and reflected upon in a performative phrase like "I hereby assert that p is true" that we can understand the meaning of the strange predicate " . . . is true" as a predicate on the level of a meta-language. For the predicate "is true" is in fact, *redundant* as long as our truth-claims are not problematized and hence are only implicitly expressed by our propositional statements. But as soon as our truth-claims are called into question, it becomes clear that the predicate "is true" is not redundant but is indispensable for argumentation. For it is only by the predicate "is true" as a predicate on the level of a metalanguage that the *self-reflective truth-claim* of our statements can be made the topic (subject) of disputation or confirmation by argumentative discourse. But this means at the same time that *argumentation* is not based on *propositions* alone but on the human competence of *reflectively proposing propositions* by speech-acts which may be made explicit by performatives.

Now man's competence for reflectively proposing propositions by speech-acts is only part of his *communicative competence;* this means that the self-reflection of his speech-acts that is the pre-condition of his truth-claims is itself made possible by the intersubjective reciprocity of symbolic interaction between men as co-subjects of actions. This situation was well understood by Hegel and G.H. Mead, but it is only the analysis of the performative-propositional double-structure of speech-acts that clearly shows what the relationship between argumentation and communicative understanding consists in. For by such an analysis it becomes clear that argumentative discourse is not constituted only by understanding the meaning of propositions, as is suggested by the semantically interpreted calculi of formal logic, but by understanding propositions and a complementary understanding of the meaning of the communicative acts through which the propositions are proposed or, for that matter, called into question, denied or corroborated, etc.

It is by reflection on this *complementary understanding of communicative acts* that one may realize why or in what respect not only science but any type of arguing presupposes those rules or norms that make up the common presuppositions of hermeneutic and ethical rationality. For, in order to understand the different actual communicative acts, one must have already understood and acknowledged those *universal pragmatic* rules that constitute the conditions of the possibility of their illocutionary effects, say e.g. the illocutionary effects of questions and assertions. And insofar as these rules are the conditions of the possibility of communicative understanding in every argumentative discourse, one may call them *transcendental-pragmatic* or *transcendental-hermeneutic* rules in a very radical sense, since they cannot be denied without actual self-contradiction.

Among these fundamental rules there are also ethical norms that must have been implicitly acknowledged. One may sum them up as the norms of an ideal speech-situation or an ideal communication-community that is necessarily though more or less counterfactually anticipated in any serious argumentative discourse. Among them there is of course the norm of reciprocal acknowledgment of persons as equal partners and hence the norm of equal rights and duties in using argumentative speech-acts for proposing, defending, explicating and possibly questioning validity-claims, as e.g. truth-claims and (ethical) rightness-claims.

Hence the most important point is that these ethical norms, which must have been acknowledged in every serious argumentative discourse, constitute the *meta-norm* for a discursive settlement of all possible conflicts about norms, i.e. the obligation, if possible, to join a practical discourse which should try to ground or legitimate concrete norms according to the regulative idea of mediating the interests of all affected people in as far as these interests constitute claims that can be defended by reasonable arguments under the boundary-conditions of concrete situations. I cannot, of course, go into the details—and, no doubt, difficulties—of an ethics of practical discourse in the present context. But I want at least to point out that the concept of an ethics as it is *presupposed* in consensual communication and hence in argumentation must not be reduced to the concept of a special ethics just for verbal discourse, such that it could be conceived of as containing just the rules of a special game or institution among other possible forms of life. For this also belongs to the very normative pre-conditions of a serious argumentative discourse that one presupposes that it is the only possible legitimation instancy for all other human forms of life, i.e. for all forms of human interaction, in as far as they are to continue consensual communication instead of settling conflicts by diplomatic negotiations or, eventually, by open strife.

NOTES

[1]Cf. Max Weber, "Über einige Kategorien der verstehenden Soziologie," in *Logos,* IV (1913) pp. 253-94; repr. in: *The Methodology of the Social Sciences,* Glencoe/Ill., 1949. Cf. D. Henrich, *Die Einheit der Wissenschaftslehre Max Webers,* Tübingen: Mohr 1952. For "Max Weber on Verstehen" cf. also F.R. Dallmayr/T.A. McCarthy (eds.): *Understanding and Social Inquiry,* Univ. of Notre Dame Press, 1977, part One. Cf. also K.-O. Apel, *Die 'Erklären: Verstehen'-Kontroverse in transzendentalpragmatischer Sicht,* Frankfurt a.M.: Suhrkamp 1979.

[2]Cf. especially "Die Objektivität sozialwissenschaftlicher Erkenntnis," in *Archiv für Sozialwiss. und Sozialpolitik,* 19. vol. (1904), pp. 24-87, and: "Der Sinn der Wertfreiheit der soziologischen und ökonomischen Wissenschaften," in: *Logos,* VII (1917), pp. 49-88, and: "Wissenschaft als Beruf," Vortrag 1919; All three texts were reprinted in *The Methodology of The Social Sciences,* Glencoe/Ill., 1949.

[3]Cf. "Politik als Beruf," Vortrag 1919; repr. in: *Gesammelte politische Schriften,* 2nd ed., pp. 493-548; also "Der Sinn der Wertfreiheit," loc. cit., and "Wissenschaft als Beruf," loc cit.

[4]Cf. K.-O. Apel, *Transformation der Philosophie,* Frankfurt a.M.: Suhrkamp, 1973, vol. II, pp. 358 ff. (Engl. trans. *Towards a Transformation in Philosophy,* London: Routledge & Kegan Paul, forthcoming) and K.-O. Apel, "Types of Rationality Today: The Continuum of Reason between Science and Ethics," in: Th. Geraets (ed.): *Rationality To-Day,* Ottawa Univ. Press, 1979.

[5]Cf. e.g. W. Dray, *Laws and Explanation in History,* Oxford Univ. Press 1957.

[6]Cf. also K.-O. Apel, "The Conflicts of Our Time and the Problem of Political Ethics," in F.R. Dallmayr, *From Contract to Community,* New York: Marcel Dekker 1978, pp. 81-103.

[7]Cf. K.R. Popper, *The Open Society and its Enemies,* London 1945, vol. II, pp. 231 ff.

[8]Cf. e.g. H. Albert, *Traktat über kritische Vernunft,* Tübingen: J.C.B. Mohr, 1968, pp. 78 ff.

[9]Cf. P. Feyerabend, *Against Method, Outline of an Anarchistic* Theory of Knowledge, London 1975.

[10]Cf. e.g. D. Riesman's analysis of the replacement of inner-directed by outer-directed behavior in his *The Lonely Crowd,* New Haven 1950.

[11]Cf. Th. S. Kuhn, *The Structure of Scientific Revolutions,* Univ. of Chicago Press 1962, and the subsequent debate in: I. Lakatos/A. Musgrave (eds.), *Criticism and the Growth of Knowledge,* Cambridge at the Univ. Press 1970.

[12]Cf. K.-O. Apel, "Types of Social Science in the Light of Human Interests of Knowledge," in: *Social Research,* 44/3 (1977), pp. 425-70.

[13]Cf. K.R. Popper, *The Poverty of Historicism,* London: Routledge & Kegan Paul, sec. ed. 1960, preface.

[14]Cf. Th. S. Kuhn, "Logic of Discovery or Psychology of Research," in: I. Lakatos/A. Musgrave, loc. cit., esp. pp. 21 f.

[15]Cf. I. Lakatos, "History of Science and Its Rational Reconstructions," in: *Boston Studies in the Philosophy of Science,* vol. 8, Dordrecht: Reidel Publ. Co., 1972.

[16]So much as a short reply to the proposal of H. Albert (*Plädoyer für kritischen Rationalismus,* München 1971, pp. 106 ff., and *Transzendentale Träumereien,* Hamburg: Hoffmann und Campe, pp. 48 ff.) to make *hermeneutics* scientific by bio-physiological or psycho-linguistic explanations of the processes of understanding. For a thorough treatment of these problems cf. my *Die 'Erklären: Verstehen'-Kontroverse in transzendentalpragmatischer Sicht,* loc. cit. (see note 1).

[17]See K. Popper, *Die offene Gesellschaft und ihre Feinde,* II. Band, Bern 1958, p. 293 (eng. trans. *The Open Society and its Enemies,* London 1945).

[18]See J. Royce, *The Problem of Christianity,* New York 1913, vol. II, pp. 146 ff.

[19]For the thesis of complementarity cf. my "Scientistics, Hermeneutics, and Critique of Ideology," in: *Towards a Transformation of Philosophy,* loc. cit. Cf. also K.-O. Apel, "The Apriori of Communication and the Foundation of the Humanities," in: *Man and World,* 5 (1972), pp. 3-37; shortened version repr. in F.R. Dallmayr/Th. A. McCarthy (eds.), *Understanding and Social Inquiry,* Univ. of Notre Dame Press, 1977, pp. 292-315.

[20]See my *Die 'Erklären-Verstehen'-Kontroverse in transzendental-pragmatischer Sicht,* loc. cit. (see note 1).

[21]Cf. J. Habermas, "Wahrheitstheorien," in: Fahrenbach, (ed.), *Wirklichkeit und Reflexion,* Festschrift für W. Schulz, Pfullingen: Neske 1973, pp. 211-65; further K.-O. Apel, "C.S. Peirce and the Post-Tarskyan Problem of an Adequate Explication of the Meaning of Truth," in: *The Monist,* 63:3 (1980), and "C.S. Peirce and J. Habermas' Consensus-Theory of Truth," in: *Transactions of the Peirce-Society* (forthcoming).

[22]L. Wittgenstein has—rightly, I think—suggested that the purpose of communicative understanding about *meaning* cannot be reached on a socially relevant scale if the purpose of communicative understanding about *truth* could not also be reached on a socially relevant scale. He thus has rehabilitated, in a sense, the ambiguity of the German word "Verständigung." Cf. L. Wittgenstein, *Philosophical Investigations,* I, §§247/242.

[23]Cf. K.-O. Apel, "The Apriori of Communication and the Foundation of Ethics," in: *Towards a Transformation of Philosophy,* loc. cit. (see note 4).

[24]Cf. K. Bühler, *Sprachtheorie,* Jena 1934, sec. ed. 1965. Cf. also K.-O. Apel, "Two Paradigms in the Philosophy of Language," in D. Ihde, D. Dellamer and D. Tracy: *Meaning and Interpretation.* Essays in Honor of Paul Ricoeur, (forthcoming).

[25]Cf. J. Habermas, "Wahrheitstheorien," loc. cit. (see note 21).

Meta-Criticism and Meta-Poetry: A Critique of Theoretical Anarchy

KARSTEN HARRIES
Yale University

1

If Kant could call his age an age of criticism, we seem to be entering an age of meta-criticism: criticism questions and deconstructs itself, ending in theoretical anarchy.[1]

Kant thought criticism inseparable from enlightenment. Having come of age, man insists that the traditional demands placed on him by religion, morality, and society be subjected to free and public examination. Only what passes this test is worthy of our respect. Kant already extended this critique to reason itself. But if this extension challenges the authority of reason, one has to wonder how serious such a challenge, which after all presupposes the authority of reason, can be. It is hardly surprising that reason survives the test. Even if Kant's *Critique* limits reason's domain, his faith in it not only remains unshaken, but is strengthened.

Today such faith is somewhat harder to come by. We wonder whether the judge was sufficiently unbiased. Does the authority of reason in fact survive a completely free and public examination? Such thinkers as Nietzsche and Dilthey, Heidegger and Derrida, have radicalized Kant's critical enterprise in such a way that faith in reason is undermined and criticism itself rendered suspect. Consider Nietzsche's claim that thought cannot free itself from its subjection to perspective and metaphor. If this is accepted, how is it possible to hold on to the traditional understanding of conceptual rigor and truth? Truths now become illusions, their origin forgotten; to tell the truth is to lie according to a well established convention. On this view the will to the truth hides a refusal of the finitude of the human situation, a will to deceive oneself. To free ourselves from self-deception we have to subvert the idol of objectivity that has taken the place of the dead God. Instead of communicating a false sense of mastery, what we write should reveal its own inadequacy.

We meet with the same subversive intent in the work of those literary critics who, impressed by post-phenomenological and post-structuralist thinking, have led theory into the neighborhood of poetry.[2] Ronald Barthes thus places the subversion of scientific discourse at the very center of his program. His emphasis on writing (*écriture*) attacks the traditional distinction between figurative and literal discourse and generates the demand that science become literature. Freeing itself from the letter that kills, theory reclaims the creativity of poetry. It becomes "a severe poem," as Harold Bloom calls his own criticism, although if we can speak here of poetry at all it is a curiously theoretical poetry about poetry: meta-poetry.[3] Good examples are furnished by Hillis Miller, who encouraged by such precursors as Heidegger and Derrida, uses etymologies not to arrive at the original or true meaning of a word, but rather to lead us, as he himself insists, into a labyrinth of meanings that renders interpretation unstable and equivocal and at the same time liberates it and lets it become playful and creative.[4]

By its very nature such poetic meta-criticism will invite uncritical response, be it acceptance or rejection. Its attack on literalism and its refusal of objectivity, so seductive to those who have been taught to associate literalism with one-dimensionality and objectivity with repression, has to put every would-be critic on the defensive: has his appeal to criteria not already been disallowed? And yet, meta-criticism can claim to be itself the product of reflections that have their own kind of rigor. If theory here ends in anarchy, is it not because critical reflection must undermine itself? It is this question that I want to examine in this paper.

2

I have spoken of the subversive intent behind the merger of criticism and poetry in meta-criticism and meta-poetry. Presupposed is a profound uneasiness with scientific rationality. The commitment to objectivity, so basic to our culture, is taken to be both mistaken and destructive. The view is familiar. In Nietzschean fashion Barthes speaks of "the theological idol set up by a paternalistic science." "The human sciences, belatedly formulated in the wake of bourgeois positivism," are seen by him "as the technical alibis offered by our society in order to maintain itself the faction of a theological truth proudly, and improperly, freed from language."[5] Like Heidegger, Barthes challenges the assumption that language is adequately understood as a means of communication and expression, that it can be purified until it becomes the obedient servant of thought. The technical understanding of language that is here challenged is inseparable from science. The commitment to objectivity demands that language not intrude itself. Ideally a neutral and transparent medium, it should permit itself to be forgotten by the language-user. Like clear glass, it should present things as they are, without distortion, for only in that way is objectivity guaranteed. But is such objectivity not an illusion, supported by our pride which would have us see as God sees, *sub specie aeternitatis,* free from the distortions imposed by our location in space and time? Such godlike vision, however, is denied to man. Language is not the obedient servant of thought, but its ruler. And language can never be rendered so pure as to be free of those perspectival distortions that are part of human existence. For Barthes there can be no escape from subjective prejudice.

> Every utterance implies its own subject, whether this subject be expressed in an apparently direct fashion, by the use of 'I,' or indirectly, by being referred to as 'he,' or avoided altogether by means of impersonal constructions. These are purely grammatical decoys, which do no more than vary the way in which the subject is constituted within the discourse, that is, the way he gives himself to others, theatrically or as a phantasm; they all refer therefore to forms of the imaginary. The most specious of these forms is the privative, the very one normally practiced in scientific discourse, from which the scientist excludes himself because of his concern for objectivity. What is excluded, however, is always only the 'person,' psychological, emotional or biographical, certainly not the subject. It could be said moreover that this subject is heavy with the spectacular exclusion it has imposed on its person, so that, on the discursive level—one, be it

> remembered, which cannot be avoided—objectivity is as imaginary as anything else.[6]

Barthes' remarks can be understood as representative of that radicalization of Kant's Copernican revolution which has shaped twentieth century philosophy. The key to Kant's revolution is the insight into the way the subject helps to constitute what presents itself to it. What we encounter is always refracted by our categories and forms of intuition. Theory cannot penetrate beyond phenomena. Things as they are in themselves lie beyond its reach. Kant, however, did not draw from this the conclusion that objectivity was an illusion. Quite the contrary, as already with Descartes, the turn to the subject was designed to secure our trust in objectivity. Descartes' interpretation of the self as thinking substance provides the key. That interpretation rests on a twofold reduction. First the self is made into a spectator of what is; the world is transformed into a picture for the thinking subject which thus loses its place in the world. This first reduction of being to being for a subject in no way overcomes relativism. To secure objectivity a second reduction is required. The self is thus further purified by being brought to the realization that the body and the senses are not essential to its being. In this abstracted self's presence to itself Descartes finds the paradigm of clear and distinct knowledge. Bracketing all determinations, thought here attempts to think only itself. Here we have that spectacular exclusion of the person of which Barthes is speaking.

We may well question how much content is left to such self-awareness. But we cannot deny that thought has the power to transcend the body. This power of self-transcendence makes it possible to oppose to the embodied self the idea of an angelically pure or transcendental self and we can ask: how would things present themselves to such an ideal observer? Using this idea as a measure we cannot rest content with the evidence presented to us by the senses. Is such evidence not subject to a point of view which is ours only because our body happens to be where it is? As Plato already emphasized, the senses can give us no more than appearances, subject to the makeup of our body and the accident of its location in time and space. Yet our body is not a prison. Not only can we move and thus gain different perspectives. In our imagination we can put ourselves in other places, both real and imaginary, without moving at all.

Neither sense nor imagination can escape perspective altogether. In the end they remain tied to the body. Thought, however, is not limited in this way. We can insist that knowledge free itself from all dependence on particular points of view and from the perspectival distortions that are inseparable from such dependence. The Cartesian turn to the self, with which, as Hegel says, "the education, the thinking of our age begins,"[7] has to lead to the demand for objectivity. Given that demand, all those aspects of nature that presuppose a reference to the human body and thus to a particular point of view have to be understood as mere appearance. This includes all secondary qualities. We gain genuine understanding of natural phenomena only when sights and sounds are reduced to measurable shape and movement.[8]

The devaluation of sense and imagination, and with it the devaluation of art, is a necessary corollary of the prestige our culture has granted to objectivity. It is this that allows Hegel to claim that thought and reflection have overtaken art, which is not to say that there will be no longer those who create or enjoy art, but it is to say that for us art no longer serves as a privileged vehicle of truth; we turn to art for entertainment, however rarefied and sublime it may be. To claim truth for art today implies a critique of the shape of our culture.[9]

Descartes' turn to the thinking subject also implies that devaluation of language that Barthes protests with such vehemence. A modern defender of objectivity would grant that we cannot do without language and, furthermore, that our language is not our own. Enshrined in it are taken for granted ways of seeing and interpreting things. Language carries with it its own points of view. Like the senses, it thus tends to seduce us into mistaking perspectival appearance for reality. But a modern Cartesian would insist that Nietzsche's metaphor of the prison-house of language exaggerates. In thought we can go beyond heard or seen signs to meanings which, although perhaps mediated by, are not bound to particular expressions, as the possibility of paraphrase and translation shows. Thus it is possible to oppose to concrete languages and to their perspectives a transperspectival realm of logical sense. And if particular languages, which are more the work of the imagination than of the understanding, obscure that sense, cannot thought free itself from these limitations, ascend to a clearer vision, and construct an ideal language, a *mathesis universalis,* which removes that deficiency? This is not to deny Barthes' claim that the subject is constitutive of what we know, nor that objectivity demands a discourse from which the person, not the subject, is excluded. But it is to deny that this exclusion renders objectivity in any way suspect. Has the *Critique of Pure Reason* not shown that it is precisely the turn to the transcendental subject that secures our trust in objectivity?

And yet Barthes is not alone in questioning this spectacular exclusion of the person and in insisting that the objectivity it yields is imaginary or phantastic. The challenge is given striking expression by Heidegger:

> The ideas of a 'pure "I" ' and of a 'consciousness in general' are so far from including the *a priori* character of 'actual' subjectivity that the ontological characters of Dasein's facticity and its state of Being are either passed over or not seen at all. Rejection of a 'consciousness in general' does not signify that the *a priori* is negated, any more than the positing of an idealized subject guarantees that Dasein has an *a priori* character grounded upon fact.
>
> Both the contention that there are 'eternal truths' and the jumbling together of Dasein's phenomenally grounded 'ideality' with an idealized absolute subject, belong to those residues of Christian theology within philosophical problematics which have not as yet been radically extruded.[10]

This suggests that in appealing to an idealized subject to ground objectivity Descartes drew illegitimately on his theological precursors. Kant's transcendental subject meets with the same objection. Is its constitutive power more than an illicit projection of God's creative power unto man? Because of Kant's failure to subject his understanding of the transcendental subject and, with it, of objectivity and truth to sufficiently critical attention, his Copernican revolution remained incomplete.

The charge is indeed obvious. It appears already in Herder's *Metakritik zur Kritik der reinen Vernunft* (1799), which protests both against the elision of the person and the elision of language in Kant's critical project. Against Kant Herder insists that we think with words and cannot think in any language other than our own. If those metaphysicians Kant criticizes have lost themselves in airless realms, Kant himself, Herder suggests, tries to rise even higher, losing himself in an empty merely formal transcendence. Instead of a *Critique of Pure Reason* Herder calls for a physiology of man's faculties and for a study of language as it is. Heidegger offers a succinct argument in support of this position:

> Discourse is existentially language, because that entity whose disclosedness it Articulates according to significa-

> tions, has, as its kind of Being, Being-in-the-world—a Being which has been thrown and submitted to the 'world.'[11]

This denies the traditional distinction between discourse or logos, understood as the timeless essence of language, and concrete language. Because man's being is always bound to a particular historical situation language can never be pure or innocent. Essentially the same argument challenges the distinction between the transcendental subject and the person.

All this suggests that if we are more radically critical than Kant himself and free his crucial insight from remnants of the Christian understanding of truth as founded in the creative and aperspectival vision of God, we will be forced to recognize that we have to submerge subject and logos in the world and thus subject both to time. It is unnecessary here to trace the history of this temporalization of structures to which Kant had given transcendental status, a development that leads from psychologism and historicism to meta-criticism. The attempt to renew Kant's transcendental program, which at one time seemed to have triumphed over psychologism—I am thinking especially of the Neo-Kantians, of Husserl's phenomenology, of Frege, and of the young Wittgenstein—led once more to a temporalization of the transcendental. Heidegger's *Being and Time* and Wittgenstein's *Investigations* document this process. Structuralism has undergone a similar evolution. Barthes' insistence on the autonomy of language provides this process of temporalizing transcendental structures with a logical conclusion.

With his insistence on the constitutive role of the subject Kant had imprisoned theoretical understanding in the realm established by the forms of intuition, time and space, and by the categories, thus denying it a knowledge of things in themselves. But if Kant thus denied man insight into what we can call material transcendence, he compensated him for this loss by insisting on the formal transcendence of the subject over the concrete person, of the categoreal framework over particular languages. It is this preservation of formal transcendence that lets one hesitate to use the metaphor of the prison at all. Kant's categories, if his transcendental argument succeeds, leave room for all possible experience. Objective reality, corresponding not to the person, but to the transcendental subject, thus transcends the appearances that it may present to a particular individual or language-community and provides them with a measure. On this view language has an outside. We can take steps to lift the veil of merely subjective appearance, as for instance

Copernicus did when he replaced the perspective-bound geocentric cosmology of the Middle Ages with his more objective account. But it is precisely the notion of "all possible experience," on which Kant's transcendental argument rests, that is suspect and with it the concept of objectivity. How are we to think of all possible experience? Will our attempt to do so not be limited to what we are able to think as possible? And will this in turn not reflect the limits imposed on our thinking by our cultural situation, including the language we happen to speak? Was this not Nietzsche's great insight? Appeals to developments in mathematics and physics that seem to discredit Kant's analysis of space and causality help to undermine the absolutism of his claim to have established the only possible categoreal scheme.[12] If we now bring Kant's transcendental subject down to earth and replace his categoreal scheme with concrete language, we are also forced to deny the distance that on the Kantian or the Cartesian account separates object and mere appearance, thinking and writing or speaking. With the idea of the transcendental subject that of an outside also collapses. This twofold loss lets language lose its measure and become free play. The boundary between theory and literature, between prose and poetry begins to blur. Transcendental philosophy thus seems to bear within itself the seeds of its own deconstruction. Theory itself seems to demand the anarchy of theory.

3

In spite of such considerations, the development that I have sketched needs to be questioned. How convincing is this critique of objectivity? To give some focus to this question, consider once more Heidegger's claim that the conception of the idea of a pure or transcendental "I" does violence to the human situation and rests on an illegitimate projection of attributes of the Christian God unto the self. That Kant's understanding of the transcendental subject, which provides objectivity with a foundation, is indeed indebted to the Christian conception of God, whose knowledge founds reality, cannot be denied. We can show that there is a direct connection between the transcendentals of medieval philosophy and Kant's use of the term. The German rationalists provide here the crucial link. But this fact of historical dependence in no way discredits the notion. The idea of a transcendental subject has its foundation in the self-understanding of man, who is aware of himself as occupying a particular point of view and in such awareness transcends it. To the extent that I can think the inadequacy of the perspective I am furnished by my body and its location I am already in some sense beyond the limits it imposes, capable of imagining or conceiving other possible points of view. This holds for any perspective. As a German folk song has it, thought breaks the walls and chains of any prison. This holds not only for prisons of stone, but also for those cultural and linguistic prisons that have obsessed modern philosophers. Heidegger's understanding of the ecstatic essence of man has to be expanded to accomodate the full scope of the freedom of thought. This boundless power of transcending makes it possible to oppose to the concrete, embodied I and to its perspectival vision the idea of a pure "I" that would not be bound by the accident of its spatial and temporal location; it would be beyond and therefore free from all perspectival distortions. In this sense it can be said that man looks up, out of his finite temporal situation, to eternity. This much at least is right about the traditional idea that man measures himself by looking up to God. Were it not for the possibility of transcending himself in his finitude, man could never have thought the aperspectival nature of God's creative vision.

The idea of an aperspectival approach to what is can be made the measure of what presents itself to us, perspectivally refracted by our senses and by our language. That measure demands forms of description that are not tied to a particular individual or group. The ideal of objectivity demands universality and translatability. To claim that we have understood something, but that in principle it can be expressed only in this particular language, suggests that we have not really understood at all. This is not to deny that, as Wittgenstein insists, we also use "understand" in a different sense, where we say that we have understood something precisely when we know that translation is impossible. Wittgenstein is thinking of poetry.[13] We may even place a particular value on this kind of understanding. But we cannot deny that it is possible to measure different types of discourse by the ideal of objectivity. Scientific discourse will come closer to it than poetic discourse. Thus the distinction between theory and literature, between prose and poetry, that seemed to disappear as a result of our deconstruction of transcendental philosophy is reasserted.

There is, however, an obvious objection. Suppose it is granted that the idea of the transcendental subject and with it the idea of objectivity is constitutive of the human situation. Can these ideas ever be more than mere ideas? The thought of a transcendental stand-point in no way assures that man can place himself there. Kant falsely assumes such a position when he seeks to determine the form of all possible experience. But what are the limits set to such experience? Kant's transcendental argument depends on the unshakable validity of the insight granted by the pure intuition of space and time. To question this is also to question the categories. But as the development of non-Euclidean geometry and of quantum physics shows, just as our imagination enables us to transcend the limits imposed on us by perception, so thought enables us to transcend the limits imposed by even a purified imagination. Man's power of self-transcendence suffers no absolute starting point. Even the principles of logic can be considered perspectival phenomena, as is shown by Nicolaus Cusanus' thought of a *coincidentia oppositorum,* although here the step beyond all perspectives becomes a step beyond the finite to an infinite to which no content can be given. The process of self-transcendence leaves us with an altogether empty formal transcendence.

To stop this flight to the aperspectival, which is necessarily a flight to an empty infinity that denies man certainty, would require an intuition unclouded by language and uncontaminated by perspective. But whatever attempts have been made to show that man possesses such an unmediated vision have failed. And without such vision it becomes impossible to nail down the terms of our language. We have to grant the deconstructionists that there is no master language unclouded by metaphor. Consider Descartes' paradigm of a clear and distinct idea, the insight we have into our own existence. To be sure, I cannot deny that I exist. But do I recognize myself in Descartes' thinking substance? At most it offers me the abstract form of my and any other consciousness, an abstraction that can appear so clear and distinct only because it is so empty. As soon as I grasp myself as this particular individual, existing here and now, the illusion of transparency disappears. The attempt to leave behind the body and to ascend to the angelic height of pure thought leaves only the ghostly shadow of the self.

As a matter of fact, even that shadow is less than a clear and distinct idea. Descartes' interpretation of the self as a thinking substance can be shown to rest on a quite traditional, but nevertheless questionable interpretation of being, according to which "to be" means "first of all to be as a substance." Descartes' attempt to see for himself, to free his vision from the perspective provided by what others have thought before him, proves impossible from the very beginning. No matter how hard a thinker may try, no matter how self-consciously he may go about bracketing the authority of what others have thought and written before him, in the end he cannot escape the tyranny of past writing and on further reflection what is presented as a clear and distinct idea will become hopelessly entangled in past writing. Leszek Kolakowski seems to me right when he suggests in his book on Husserl "that a truly radical search for certitude always ends with the conclusion that certitude is accessible only in immanence, that the perfect transparence of the object is to be found only when the object and subject (empirical or transcendental Ego, no matter) come to identity. This means that a certitude mediated in words is no longer certitude."[14] Again we meet with the attempt to assimilate human vision to the creative vision of God. But God does not know the gap between subject and object, word and thing. For man the dream of an immediate vision must remain just that, a dream.

And yet, we cannot escape that dream. As Sartre knows, regardless of whether we believe in the existence of God or not, we measure ourselves by the idea of God. And if divine or angelic vision is denied to us, we can at least develop forms of representation that are more objective than ordinary language and using these forms recast experience, forcing it to yield a more adequate, if not adequate knowledge. Similarly we can classify expressions by the extent to which they permit or resist substitution of some other expression without serious loss. Such a classification allows us to hold on to the distinction between theory and literature that deconstructive criticism threatens to evaporate. The regulative power of the idea of objectivity cannot be denied.

To be sure, if with Descartes we insist that knowledge rule out all conjecture and demand truth and nothing but the truth we will have to end up despairing. But there is no need to do so. How many modern sceptics are disappointed Cartesians?

4

That it is not a necessity of thought that leads to meta-criticism is suggested by its rhetoric of subversion. A need for subversion exists only when the power of what is to be subverted is both taken for granted and thought to be a threat. But what is the threat posed by the commitment to objectivity? Two charges especially recur in meta-critical literature: 1) that the prestige accorded to objectivity prevents us from doing justice to the demands of creativity and freedom, and 2) that by generating a false reality principle the commitment to objectivity lets us lose touch with nature, others, and with our own self, especially with our sexuality. I shall consider each charge in turn.

That there is a tension between the commitment to objectivity and the demands of man's being must be granted. Consider once more the reductions presupposed by the evolution of modern science. First the self is disengaged from the world and made a spectator of what is. Instead of living in the midst of things, man is placed at a distance from them, including from his own body. Things become objects for an ideally disinterested awareness, the world a picture.[15] The more complete this disengagement, the less this picture-world will assign man his place, his ethos. The spectatorial stance has to transform the world into a strange and meaningless other. Nihilism is the other side of objectivity.

But this first reduction is insufficient to establish objectivity as it is demanded by science. A further reduction is required. As has been shown, objectivity requires that we rise beyond perspectival distortion. The self is purged by being brought to the realization that the body and the senses cannot limit the power of thought. Thought is made the measure of reality: what is is not simply for the subject, but for the purified thinking subject. Only in this way can we distinguish mere appearance from objective reality. Implied is the devaluation of secondary qualities. By its very conception objective reality has to exclude them. The world has lost not only its meaning, but its color as well. The second reduction requires a form of description from which all reference to particular points of view has been eliminated. Mathematics provides that form.

At this point the charge that the commitment to objectivity estranges us from the reality of nature and from our own reality has some plausibility. The Cartesian turn to objectivity does indeed seem to transform nature into something altogether imaginary. And yet, the development of modern science makes it impossible to take seriously the charge that as a result of sacrificing practice to theory Descartes lost touch with reality. As a matter of fact, Descartes' flight beyond the world of the senses is designed to grant man power over that world. Nothing could characterize better the cognitive impulse that lies at the basis of our technological culture than Descartes' claim that his principles "opened up the possibility of finding a practical philosophy by means of which, knowing the force and action of fire, water, air, the stars, heavens, and all other bodies that environ us, as distinctly as we know the different crafts of our artisans, we can in the same way employ them in all those uses to which they are adapted and thus render ourselves the masters and possessors of nature."[16] It should be noted that the model of understanding that is provided here is not detached contemplation, but the artisan's know-how. Man comes closest to the creative knowledge of God when he knows how to make something. Purely mathematical speculation cannot give insight into nature. Descartes therefore insists on mechanical models. We understand things to the extent that we can make them. To fully understand nature is to be able to recreate it. To fully understand man is to be able to make a human being.

It is precisely because theory here is not content to remain purely theoretical but demands to be completed by practice, a practice that cannot let nature or man alone, that it constitutes a threat. Promising man a godlike position, Cartesian science serves what the tradition had called pride. Perhaps we should rather speak of a will to power—the term may be preferable in that it does not imply criticism. But then, can we afford not to be critical? Must we not ask to what extent science and technology leave room for what is essentially human? How, for example, are we to reconcile our conviction that persons demand respect with the refusal of science and technology to accept boundaries? Is such respect compatible with human engineering? If we must grant that man can transcend himself as a concrete person and thus free himself from perspectival distortions, does not the more objective approach made possible by this flight obscure the person? That this ascent grants us power cannot be denied. Nor that it has shaped the world we live in and our common sense. Because of this the common attempt to escape from the Cartesian shadow by appealing to ordinary language or to the life-world is necessarily ineffective. They themselves are heavy with our Cartesian inheritance. But if ineffective the attempt shows that much of the confidence in the compatibility of the demands of science and those of humanity is now gone. What place do science and technology leave for creativity. Man's will to power threatens to overpower his own essence.

In *Notes from the Underground* Dostoevsky suggests that if science were to provide man with the keys to nature, including his own nature, so that all could be calculated and no place would be left for chaos and darkness, unwilling to deny his own freedom man would seek refuge in madness. The more we succeed in satisfying our desire for form and order, and the closer we come to fulfilling the Cartesian dream of rendering ourselves the masters and possessors of nature, the more passionately we will pursue anarchy and the mysteries of the labyrinth.

It may be objected that such a despairing response rests on a misunderstanding of science that not only conflates science with natural science, but falsely absolutizes the claims of the latter and illegitimately extends its domain. Have we not learned that science can offer us no more than conjectures? And did not Descartes already succeed in showing that mechanical models cannot in principle do justice to the being of man, and this for the simple reason that a science that excludes final causes prevents itself from understanding what is distinctively human? The limitations of mechanical models were stated with all possible clarity by Heinrich Hertz in his preface to *The Principles of Mechanics:*

> The same sense that leads us to eliminate every trace of an intention, of pleasure and pain, from the mechanics of the inanimate world as foreign to it, lets us have misgivings about depriving our picture of the world of these richer and more varied representations. Our fundamental law, perhaps sufficient to represent the motion of inanimate matter, appears at least on brief consideration, too simple and limited to represent even the lowest life processes. That this is so seems to me not so much a disadvantage as an advantage of our law. Just because it enables us to gain a complete view of the totality of mechanics it shows us the limits of that totality.[17]

Not only did Hertz recognize that the model he had proposed was only that—a model, and by no means the only such model that could do justice to the phenomena—but also that the reductions of experience presupposed by the form of description chosen prevented it from doing justice to the different dimensions of reality, especially to the human dimension, and he took it to be a special virtue of his presentation that it made the nature of these reductions so clear that they could not be overlooked. With his mechanical models man does not seize reality as it is. Hertz emphasizes the fact that nothing in the world corresponds to his material points; they are ideal constructs required by the model. Yet their merely ideal status in no way robs the model of its predictive power.

This raises the question whether the failure of Hertz's model to do justice to human reality has its foundation in the particular type of model he had chosen or whether it attaches to the commitment to objectivity itself, a commitment that we may take to be inseparable from science. Can there be a science of human reality that does not elide man's being as a person, his freedom and his creativity?

This may seem a curious question. Are there not the human sciences? We can point to Schutz's attempt to use Husserl's phenomenology to establish the scientific character of the sciences of man or to structuralism. Both promise interpretations of human behavior that do not rely on efficient causality and yet possess the required objectivity. But can a purely objective thinking grasp the meaning of human behavior? Man's actions and works can be understood only in terms of the necessarily subjective intention or project that governs them. In spite of Schutz's claim to have done justice to both the demand for objectivity and to the essential subjectivity of meaning, the two cannot finally be reconciled.[18] Schutz thought such a reconciliation was effected by Weber's ideal types, which are described as constructs arrived at "by postulating certain motives as fixed and invariant within the range of variation of the actual self-interpretation in which the Ego interprets its own actions or its acts."[19] Ideal types are thus idealizing objectifications that have their ground in self-understanding. But this is a shifting and uncertain ground. Not only are we not transparent to ourselves, but the subject's self-understanding will depend on his historical situation. The models provided by the construction of ideal types are thus essentially perspectival.

One could object that Schutz preserves the objectivity of the human sciences with his insistence that, like mechanical models, the constructs of the human sciences be "causally adequate," i.e. that they should "predict what actually happens in accord with all the rules of frequency."[20] It is difficult to object to this demand, although it leaves one wondering why the social scientist should not content himself here with talk about the average behavior of a certain group of actors. Schutz would insist that such understanding from without is deficient in that it does not give us any insight into the meaning of what is being observed.

I do not want to deny that the interpretive sociology of Weber and Schutz is better able to do justice to the human situation than a sociology that pays attention only to the external course of action. This is the case, however, precisely because in order to remain in touch with the meanings of our life, it is willing to relax the requirement of objectivity. Dilthey here proves a better guide than Husserl. Dilthey's understanding of life and history has to challenge all claims to objectivity and universality in the human sciences and invites theory to learn from poetry. His attempt to elevate hermeneutics to *the* method of the human sciences, and its development by Heidegger and Gadamer, acknowledges the necessity of freeing the human sciences from what

Gadamer calls "the ontological obstructions of the scientific concept of objectivity."[21] This denies that the human sciences can or should develop models that possess the kind of objectivity that Hertz sought to guarantee by his form of description. In the human sciences at least the insistence on objectivity must be questioned. Where it goes unquestioned it must do violence to man. Barthes' description of the human sciences as the technical alibis which our culture uses to endow what it would have us do with an aura of necessity is not easily dismissed. It is in this context that Geoffrey Hartmann's claim that "Anything that blows the cover of reified or superobjective thinking is important" can be granted.[22] But if it is important to put the commitment to objectivity in its place, especially so in the human sciences, this should not lead us to deny it a place. The human sciences have to preserve the tension between the requirements of objectivity and subjectivity if they are not to turn into anemic poetry. In the human sciences, too, there are facts to be ascertained and lawlike structures to be discovered, even if we have to insist that this is not enough and that we have to go further if we are to understand the meaning of human actions and works.

Meta-criticism refuses the tension. This suggests that its subversion of superobjective thinking aims at more than just to put objective thinking in its place.

5

In his *Lectures on Aesthetics* Hegel claims that "Thought and reflection have overtaken the fine arts." It is difficult to disagree. If we grant Hegel that art in its highest sense is more than sublime or perhaps not so sublime entertainment, but that it has its place "in the same circle with religion and philosophy" as one way of articulating the most profound truths of the spirit and if, at the same time, we agree with him that the modern situation is shaped by the insight that truth demands objectivity, then his conclusion becomes inescapable. Given such an understanding of truth, the locus of truth can alone be thought. Whatever is presented to our senses must then first be translated into the medium of thought if it is to yield truth. This is also necessary if we are to find truth in art.[23]

In support of Hegel one could point out that increasingly works of art are seen as little more than occasions for endless reflections. The critic has begun to rival or even to replace the artist. Meta-criticism is a case in point. But while the phenomenon of meta-criticism lends some support to Hegel's thesis, it also forces us to question it: When Hegel speaks of thought and reflection he is thinking of scientific understanding. But meta-criticism does not pretend to offer such understanding. Quite the contrary: Refusing the boundary that separates theory and poetry, refusing to subordinate the imagination to reason, the phenomenon of meta-criticism suggests that if thought and reflection have indeed overtaken the fine arts, they in turn are being overtaken by a higher art. Theory turns poetic, becomes meta-poetry. This turn presents itself to us as not just another aesthetic escape from reality, but as a challenge to the prevailing reality-principle. Inseparable from meta-criticism is the attack on the logo-centrism of our culture, which leaves only a peripheral role to poetic vision and imagination and would have us deny our sensual and sexual being. On this point there is agreement between Norman O. Brown and Marcuse, Barthes and Derrida, Geoffrey Hartmann and Hillis Miller. It links meta-criticism to a tradition that includes surrealism and has one origin in the magic idealism of Novalis.

Common to the meta-critics is also the attempt to interpret this logocentrism by means of Freudian metaphors. The characterization of logocentrism as phallocentrism challenges the purity that theory had claimed for itself. The autonomy of theory is denied as logos is interpreted as a phenomenon of eros. Meta-criticism thus attempts to recover the erotic character of writing, which logocentric theorizing represses or denies. It is easy to be seduced by this rhetoric which seems to have given philosophy and literally criticism, which not so long ago seemed to be on the verge of expiring from anemia, new life and interest. How interesting to learn that "The element of play in language is the erotic element," especially so when this element is said to be "in essence not genital, but polymorphously perverse."[24] Or to have writing compared to masturbation.[25] And what possibilities for further play are opened up by Barthes' discussion of the letter Z as the letter of mutilation and castration: "phonetically Z stings like a chastising lash, an avenging insect; graphically, cast slantwise by the hand across the blank regularity of the page, amid the curves of the alphabet, like an oblique and illicit blade, it cuts, slashes, . . . "[26] Even from a simple letter the imagination wrests drama, soaring beyond common sense. And yet, in spite of all their Freudian imagery, there is very little eros in meta-critical texts. Eros is only evoked from a great distance. The desired recovery of the erotic remains abstract and unreal. What we have is poetry in a theoretical medium, the work of a disembodied imagination, no longer capable of healing the breach between eros and logos.

Even as it longs for reality, meta-poetry refuses it. The subversion of the ruling reality-principle remains ineffective because merely aesthetic.

Striking is not only the prevalence of sexual imgery in meta-criticism but its preference for the abnormal. Instead of the dominant phallocentrism we have a celebration of the polymorphously perverse, of narcissism and masturbation, of "fantasies too thrilling to be actualized or ended."[27] This preference for the exorbitant and perverse may be understood in terms of the pursuit of the interesting. After all, as Friedrich Schlegel and Kierkegaard insisted, the abnormal will always be more interesting than the normal. But such an interpretation of the auto-erotic aspect of meta-criticism, while convincing, does not go far enough. Auto-eroticism implies a denial of the provocative distance that separates us from a loved person. This denial is closely related to the attack on the polarity of subject and object that is central to the meta-critical project. A corollary of that attack is the blurring of the boundary between reader and author, between interpretation and text. Marcuse's analysis of narcissism as "an attempt to engulf the environment, to integrate the narcissistic ego with the objective world,"[28] describes the meta-critical project if, instead of "environment" and "world" we read "texts." It should be noted how close, despite the obvious difference and the anti-Cartesian rhetoric of meta-criticism, this project is to the Cartesian project of rendering man the master and possessor of nature.

It is this affinity that allows Valéry to interpret Descartes as the precursor of today's meta-poets and meta-critics. According to Valéry with the *cogito* Descartes' will to power calls him out of the world, back to himself, "to his uniquely personal mission, his unshared destiny," even if the pursuit of that destiny must make him "deaf, blind, unfeeling towards everything—even towards truths, even towards realities—that may cross his career, his fated achievement, his line of growth, his inner light, his orbit."[29] Descartes would have had some difficulty recognizing himself in this description. He could have pointed out that his turn to the thinking self was not a turn to what is uniquely personal, but subjected the person to what is objective and universal; and that far from eliding nature the insistence on mechanical models acknowledges its rule and thereby prevents the spirit from losing itself in idle constructions. But if Valéry distorts Descartes' position, this distortion lets us see what it is that leads theory to take an aesthetic turn. It is precisely Descartes' pride, raised to a still higher level by Valéry's interpretation. The insistence to be, like God, sole author of one's thoughts must lead to a refusal of outside reality and to an elision of the body.

The dream of a self-sufficient, unmediated vision that has always haunted man reappears, now transformed into the dream of a language that denies the distance between itself and its object, a language that takes itself for its object and thus becomes autonomous.

Given what meta-criticism has drawn from Nietzsche, it seems appropriate to conclude with a sentence from *Zarathustra:* Wenn die Macht gnädig wird und hinabkommt ins Sichtbare: Schönheit heisse ich solches Herabkommen."[30] Man wills power and is yet cast into the world, dependent, vulnerable, and mortal. Unable to forgive himself his lack of power, man's will to power turns vengeful and rancorous, pale and sublime. It is this spirit of revenge that denies man beauty, where beauty should not be understood aesthetically, but as the epiphany of the real. As Nietzsche suggests, the spirit of revenge can be overcome only when power becomes gracious, that is to say, when man learns to forgive himself his lack of power and thus gains the strength to accept himself as he is, in his finitude and insufficiency, but also in his divinity. Without such acceptance there can be no openness to what transcends him and alone can give him his measure. In the spirit of revenge we find the origin of theoretical anarchy. But the spirit of revenge is nothing other than what the tradition had called pride. As Augustine knew, when the soul becomes a kind of end to itself" it grows "frigid and benighted."[31]

NOTES

[1]The term "theoretical anarchy" was suggested to me by Reiner Schürmann's "Political Thinking in Heidegger," *Social Research,* vol. 45, no. 1, p. 214.

[2]I am thinking especially of the critics Geoffrey Hartmann calls "revisionist," "hermeneutic," or "post-New Critical." See "Literary Criticism and Its Discontents," *Critical Inquiry,* vol. 3, no. 2, pp. 203-220.

[3]Harold Bloom, *The Anxiety of Influence. A Theory of Poetry* (New York, 1966), p. 13.

[4]J. Hillis Miller, "Ariadne's Thread: Repetition and the Narrative Line," *Critical Inquiry,* vol. 3, no. 1, p. 70.

[5]Roland Barthes, "Science versus Literature," *Introduction to Structuralism,* ed. and int. Michael Lane (New York, 1971), pp. 415, 416.

[6]*Ibid.*, p. 414.

[7]*Vorlesungen über die Geschichte der Philosophie, Jubiläumsausgabe,* ed. Hermann Glockner, vol. 19, p. 329.

[8]See my "Descartes, Perspective, and the Angelic Eye," *Yale French Studies,* no. 49, pp. 28-42.

[9]See my "Hegel on the Future of Art," *The Review of Metaphysics,* vol. 27, no. 4, pp. 677-696.

[10]*Being and Time,* tr. John Macquarrie and Edward Robinson (New York and Evanston, 1962), p. 272.

[11]*Ibid.*, p. 204.

[12]See Stephan Körner, *Categorial Frameworks* (Oxford, 1970), pp. 63-74.

[13]*Philosophical Investigations,* tr. G.E.M. Anscombe (New York, 1953), p. 83.

[14]Leszek Kolakowski, *Husserl and the Search for Certitude* (New Haven and London, 1975), p. 83.

[15]See Martin Heidegger, "Die Zeit des Weltbildes," *Holzwege* (Frankfurt am Main, 1950), pp. 69-104.

[16]*Discourse on Method VI,* in *The Philosophical Works,* tr. Elizabeth S. Haldane and G.R.T. Ross (New York, 1955), vol. 1, p. 119.

[17]Heinrich Hertz, *Die Prinzipien der Mechanik* (Leipzig, 1894), p. 45.

[18]See my review "Alfred Schutz, *The Phenomenology of the Social World,*" *The Journal of Value Inquiry,* vol. 4, no. 1, pp. 65-75.

[19]Alfred Schutz, *The Phenomenology of the Social World,* tr. George Walsh and Frederick Lehnert (Evanston, 1967), p. 244.

[21]Hans Georg Gadamer, *Truth and Method* (New York, 1975), p. 235.

[22]Geoffrey Hartmann, "Literary Criticism and Its Discontents," p. 216.

[23]See Hegel, *Vorlesungen über die Aesthetik, Jubiläumsausgabe,* vol. 12, pp. 31-32.

[24]Norman O. Brown, *Life Against Death. The Psychoanalytical Meaning of History* (Middletown, 1959), p. 70.

[25]Jacques Derrida, *Of Grammatology,* tr. Gayatri Chakravorty Spivak (Baltimore, 1974), pp. 141-164.

[26]Roland Barthes, *S/Z. An Essay,* tr. Richard Miller (New York, 1974), p. 106.

[27]Richard Rorty, "Derrida on Language, Being, and Abnormal Philosophy," *The Journal of Philosophy,* vol. 74, no. 11, p. 681.

[28]Herbert Marcuse, *Eros and Civilization* (New York, 1962), p. 153.

[29]Introductory essay to *The Living Thoughts of Descartes,* tr. Harry Lorrin Binsse (London, 1948), p. 31.

[30]*Zarathustra, Zweiter Teil,* "Von den Erhabenen."

[31]St. Augustine, *The City of God,* XIV, 13, tr. Marcus Dods (New York, 1950), p. 460.

Practical Philosophy as a Model of the Human Sciences

HANS-GEORG GADAMER
University of Heidelberg

The title of my talk today could serve as a heading for this whole conference. It is evident that the expression "human sciences" is problematic for us today and that we must come to the conclusion that science should be defined by us in another way than it is for modern times. This, of course, includes a certain justification of the older Greek conception of knowledge as "philosophy." It is not because of my special predilection for the Greeks that I propose this topic for today, but rather because of the necessity of seeking an epistemological self-understanding which is not based on the credence of the natural sciences and of the ideal of method as it was characteristically called at the beginning of the seventeenth century and as it dominates the research work and our academic activities in the humanities. It is for this reason alone that I want to go back to the philosophy of Aristotle, for it is this ancient philosopher who defended for the first time a special approach to the subjects of human action and human institutions.

That it makes sense to investigate from this point of view the methodological impasses of our contemporary situation seems to me quite well illustrated by the meeting for which we have gathered together today. We have seen how the problem has already been expressed by Professor Apel in connecting it with this almost doctrine-like radicalism of Max Weber and the whole question of value-free research which serves to verify the distinction of public and private and the orientation of anything that is public to the model of what Husserl called "knowing for everybody." To this position we must ask: Is this not the expression of the conflict in our own self-understanding rather than the solution to this problem? No humanist, no follower of the sciences that we call *Geisteswissenschaften,* can be described in his own doing and his own ideals under the point of view of value-free science. That is so self-evident that I find it necessary to reconsider the whole question of the humanistic tradition and its function in our modern culture.

It is certainly a very ambiguous situation in which the humanities have continued their activities since the beginning of modern humanism in the fifteenth and sixteenth centuries. It is not very comfortable to choose between the rhetorical tradition of the *artes liberales* or to side for a self-understanding that calls itself critical and methodical and tries to compete with the natural sciences. I am referring here to John Stuart Mill's well-known chapter in *A System of Logic* in which a naturalistic methodological ideal is assigned to the human sciences. It ends by saying that the "moral sciences" (*Geisteswissenschaften*) constitute a field of modern sciences which can compete with the prognostic exactitude of meteorology. I think under this point of view the humanities have a very poor place and a very limited self-understanding compared with the fact that the contents which are treated and interpreted by the humanities have such a tremendous impact on the whole world view, on the whole moral attitude and on the expectations of society.

Confronted with this modern ideal of methodical sciences, we cannot avoid seeking for the enterprise of reason in the humanities a better recognition than that of the procedures of modern science. Far from recommending something as a new method, I can but describe how the critical, methodical attitudes of modern investigation cooperate with another human tradition. To this end I proposed to resume the expression hermeneutics, which, of course, I take from Dilthey and Heidegger, and which is connected with the whole problematic of so-called "historicism"—a philosophical situation which should also inform my own contribution to the subject.

It is obvious that between Dilthey and Heidegger we have a very radical gap, even considering Heidegger's statement that he will continue in elaborating the scientific thoughts of Dilthey and of his philosophical friend Graf Yorck. Even when he had said that, the fact remains that Heidegger in *Being and Time* had called his own dimension of questioning a hermeneutics of facticity. Dilthey, as one of the great admirers of the positive sciences of the nineteenth century, one who was a dedicated admirer of Hermann von Helmholtz, the greatest figure of the University of Berlin at the time he accepted the call to Berlin, was nevertheless overwhelmed by the breadth and richness of the idealistic heritage. Hermeneutics, in his own eyes, was in the end the deciphering of the ideal book, the book of history, the book which was written by human culture in its ongoing process of authorship. So the new Heideggerian expression "hermeneutics of facticity" formulated a real challenge. If we examine the word "facticity," we see that "*Faktizität*" in German is no less artificial than "facticity" in English. This word is not a neologism of Heidegger, rather, it is an expression which originated in the theological discussions about the historicity of Jesus Christ and the resurrection.

Of course, it is true that "fact" can be defined through the opposition to meaning, significance, interpretation, or theories as results of investigation. But to say that a fact is just a pure fact is, concerning its meaningfulness, the zero point of meaningfulness. It is not the fact but just the context which defines the meaning and the significance of a fact. When neo-Kantianism, the philosophical school which helped Max Weber to define his own enterprise, speaks about value relations in the humanities, then they raised the question: What is an *historical* fact? It is obvious that an historical fact is not, in the first place, merely something that really happened, but rather, something that really happened in such a way that it has a special signification for an historical question, an historical context.

Of course, you may say that it is the same in the natural sciences. Indeed, in the natural sciences the dimension of hermeneutics—i.e., of interpreting the context which makes something meaningful—is now more or less accepted. We know that it is not sufficient to say that the natural sciences have to do with experimental facts and the humanities with interpreted facts. We know that the concept of fact in the experimental sciences involves a hermeneutical dimension because the pure increasing and augmenting of our experimental data does not constitute advancement in scientific research work. It is the *question* which defines the meaningfulness of experimental data, and they, by answering a question, are "facts."

To underline the concept of fact and facticity in the way in which Heidegger did makes us aware of the limitation of understanding and self-understanding in such a way that the idealistic heritage—this ultimate and highest ideal of Hegel to advance and reach the full transparence of the object by its own merging with the intelligence of the thinking subject—can not adequately serve as a basis for the task of self-understanding. The whole idea of the philosophy of identity was no longer a sufficient basis for the new enterprise. It began with the re-examination of Max Weber's methodology of the sciences. This great hero of my own youth, this man with his extreme radicalism insisted on an ideal of rationality which restricted scientific rationality to the questioning and seeking of answers in the technical realm of the relationship of means and ends.

And so the new attempt to introduce the limitation of self-understanding as the ontological structure of human *Existenz* becomes a decisive event. But we must not be misled by the application of the false rubric of irrationalism to this inquiry into human existence. It is to the French philosophers of our century that we owe this stigma of existentialism, and it is this situation, we can say, which serves to underscore the radical enterprise of a thinker like Heidegger and thus opens new ways for a better self-understanding of the humanities. We should recall here the leading ontological concepts with which Heidegger deals in his well-known explanation of *Vorhandenheit, Zuhandenheit,* and *Dasein*. The English translations ("presence-at-hand," "readiness-to-hand") are not acceptable to me. I cannot follow them because the German ear cannot hear in *Vorhandenheit* the *Hand* at all. And you, in the artificial translation "presence-at-hand" must of course realize the *"Hand"* in this context. As a matter of fact, *Vorhandenheit* is the opposite of hand, it is the dimension of data constituted by the experiment and by the measuring in the natural sciences which was formulated by Heidegger's expression; and his point of view was that objectivity and objectifying thinking is based on the dominating understanding of Being as "presence" in Western civilization.

Initially, I did not realize that the practical philosophy of the Greeks could give us a new self-understanding under the point of view of science. But I think that at least Aristotle offers us a better understanding of human life than can modern science, so that, for example, we do not have to accept the calculation of the validity of historical research work, of history, under the point of view of methodical critical science as reasonable. One of the members of the circle of Vienna, Victor Kraft, in his book about the historical sciences, said that at best there is ten percent science in the work of the historian, and I think that is in a way quite adequate. But it is inadequate in the sense that ninety percent would be poetry or fiction. These are not exclusive alternatives.

That is self-evident. Thus, the question for us is: How can we develop a concept of knowledge and science which really corresponds with what everyone is doing in the humanities? I cannot help but repeat again how, with respect to my own work, I was very far from introducing a new method in the humanities. It is quite evident that one must learn method to do the work of a humanist. I am, I hope, a good interpreter and a philologist. As a thinker I just wanted to propose a better understanding of what we are doing in the humanities; and perhaps that goes even further than do the humanities in the narrow sense of the word and encompasses the "sciences" as well. I must also recognize here the bold impetus for this direction opened by the phenomenological approach of Husserl. The success of words is perhaps the most reliable test for a real turning in thinking and so in the case of Husserl we have one such word, "life-world," "*Lebenswelt.*" This word is no mere neologism but as a term this word expresses the way in which Husserl attempted to expand the whole phenomenological approach beyond the limitations of scientific activities and to study the ordinary experience of life. So one of the standard achievements was his description of the structures of sense perception—for example, that we can never see at the very same moment the front and back of the sensuous object.

But how does the philosophy of Aristotle lend itself to this discussion? How can the philosophical analysis of human life and human attitudes, human actions and human institutions by the ancient thinker contribute to better understanding of what we are doing in the humanities? To ask such a question leads us to a certain re-understanding of the ongoing tradition of practical philosophy and rhetoric throughout the centuries. In reference to politics, in particular, I should mention that it was only in the post-Romantic era of the nineteenth century that politics ceased to be the title of a philosophical and methodical discipline. Since the beginning of this century we speak of "political science"; but I am not at all convinced that the scientific realm of political philosophy has been extended under this new rubric. Perhaps the older understanding of politics had a deeper relevance than we imagine.

When we turn to the origins of this tradition in Aristotle we see that it was within a special situation of the Platonic academy that he developed his own position. Aristotle, the son of a physician, was oriented in his approach towards the phenomenon of life and to the model of movement. The order of life as well as the order of the universe follow the model of the living organism. To be sure, on this point he was not absolutely in opposition to the new Platonic project of a dialectical elucidation of reality; but obviously there was a decisive difference between Plato and Aristotle regarding one crucial point, namely, the prevalence of mathematics in Plato, in contrast to the predominance of the model of the living body, of life, in the approach of Aristotle.

It is not my concern here to speak about the common ground which Aristotle shares with Plato, especially on this question of the good, but merely to point out this difference which separates their respective projects. In part, we see this difference in the question of universals which was so sharply debated by Aristotle in his criticism of the theory of forms. Nevertheless, the *Physics* is an elaboration of the Platonic program broached by Socrates' statement that he would not be content that he had understood reality as long as he failed to see the universe as being ordered the way it is because it is good for it to be so ordered. It is an anthropological and anthropomorphic model which governed Plato's as well as Aristotle's approach to physics. That is well-known. It is also well-known that Aristotle was not a founder of a system. We must realize that the concept of system is a very late epistemological concept which begins in the seventeenth century. It is provoked by the new conflict about which we are speaking, i.e., that the sciences claim to be the only reliable approach for an investigation into reality in such a way that the metaphysical tradition of two thousand years was suddenly deprived of its own legitimation. I think what we call a system in philosophy is just the attempt to bring together the contradictory approaches of the metaphysical tradition and of the new sciences.

Although we do not speak of a system in Aristotle, his genius is unmistakable. In approaching human affairs he tried to describe what happens in human life and how we can clarity the foundations of our attitudes and of our institutions. And although there are indeed problems concerning the relationship between the *Ethics* and the *Politics,* it is not my concern here to *describe* the problems of these books to you. Yet there is one point that should be brought to our attention, namely, that no possible use of the words "public" and "private" is adequate to describe the difference between the subject matters of these two books. Both concern the public and both concern the life of the person as an individual. It is not my intention to recount how Aristotle describes the constitution of the city, the system of education and institutions or the formation of our ideas of moral virtue. Nor is it relevant here to see how Aristotle takes up the Socratic-Platonic notion that knowledge, giving an account, giving a justification by reason, is the decisive factor in the life of human society. I want, rather, to focus on the question: In what sense do these books about ethics and politics present us with a science? And along with this there is a further question: How did Aristotle understand his own activity as a teacher in this field? This second question is not only Greek but almost contemporary, because it is related to the demand that philosophizing about ethical matters should have an ethical relevancy. This, seems to me to be a demand one cannot escape, even though we see the opposite in the analytic trend of British philosophy. But with thinkers like Kant or Hegel or Husserl or Scheler, one has to realize that their theoretical investigations in the field of practical philosophy will not only serve to advance theoretical clarity but must also have some impact on the clarity of decision-making and on our general orientation in practical and political life.

That is a problem which indeed occupies Aristotle as well as all the other major figures. It is for Aristotle quite clear that to investigate the basis of the good life, of happiness, of virtue, of practical reason, should also contribute something to the goodness of human life. That is a very hard thing to accomplish. It is somewhat curious that the work done in the classroom claims to have an impact on our own practical life. Aristotle is well aware of this problem; that is, one of his presuppositions—and I think it is a very reasonable presupposition—was that the formation of man by education must be done before one can begin to theorize about the practical attitudes and institutions of human life. Of course that does not mean that it is a certain unchangeable formation which is given by education. In any case, the main point is that theorizing about practical matters always involves a special risk—the risk that through this theorizing, which is always connected with a generalizing, the special case, the particular situation, comes to be seen in the light of general rules, so as to lose somewhat its own urgency and its own moral challenge. This inner tendency of our reason toward theorizing in surpassing our practical situations of action is deeply rooted in our capacity to distantiate everything linguistically.

This was, I think, one of the deepest and profoundest insights of Plato, who never had the illusion that dialectic is able to overcome once and for all the emptiness of sophistic argumentations. In Aristotle too there is deep insight into the risk which is involved in theorizing, at least in the field of practice. It is evident that human beings apply their reason for fulfilling the demands and claims which are raised by the society in which they are living. And it is also quite evident that reason offers not only the confirmation for the validity of these prescriptions, not only the insight and the means to fulfill the order of society, but also that it has the other possibility of distantiating from the moral and social rigour of the valid patterns.

And so the whole question of the relationship between morality and reason is problematic for us. With respect to the tradition which tries to mediate this relationship, I think it was Kant who had the deepest insight. I regret that in America the moral philosophy of Kant is not appreciated in an adequate form and misunderstood as a morality of blind obedience to "duty." I would like to take this opportunity to make one point about Kant's moral philosophy as it concerns the function of reason in practical efforts. Kant was deeply influenced by Rousseau. There is a note in which Kant wrote *"Rousseau hat mich zurecht gebracht,"*[1] "Rousseau corrected me." His reference was to the ambition of the age of Enlightenment, the belief that by advancement of our civilization and of our scientific progresses culture and society are improved. For Rousseau there is only corruption, not progress, to be gotten by civilization and it was through this challenge to the Enlightenment that Rousseau "corrected" Kant. In his moral philosophy Kant tells us that the sensitiveness of moral conscience cannot be advanced by any philosophical efforts. The simple ordinary man who is honest and feels what his duties are has the highest degree of conscience and of reasoning. "It is a charming thing, this man of innocence, but the innocence of a man who knows what his duty is, is so easily corrupted." It is easily corrupted by what Kant calls the dialectic of practical reason; by this he means that practical reason, the reason in practical efforts, is always eager to lower the demands of the moral and political claims; i.e., to find a new argument for the special situation in which one finds oneself so that it would be justifiable to lower these claims and thus find an escape from the special obligation. We know the answer which Kant gave to this dialectic of practical reason—viz., the categorial imperative. This famous formulation serves to suppress the sophistic of our passions. It is nothing else than the demonstration that the exception to the law cannot be admitted by reason.

Let us return to Aristotle, because as a matter of fact I think we should learn from both, for the same problem occurs in Aristotle. When Aristotle maintains that education should not be done in the classroom, but should be presupposed, he means that one should not be educated in an intellectual way so that one would pretend superiority in a moral sense. This presupposition of Aristotle is in a way quite adequate to the only description of moral reasoning in his analysis. I am referring here to *phronesis,* this habit of practical reasoning, this virtue which preserves us against corruption by emotional impacts. On the one hand, this practical reasoning corresponds to Max Weber's idea of finding the right means to ends. But on the other hand practical reason must give account of the end itself and of why we have to prefer something to something else; and, so far as we do so, we are not blind to prescriptions. We try to find the best, the good, in our decisions and this is always a very concrete thing. And not at all a blind obedience to the prescriptions of a society. It may be the case that it is good and moral, even politically justified, to go beyond the prescriptions of conventions.

I should like to add here a remark about theorizing with respect to practical behavior. It is well-known that Greek epistemology or philosophy of science did not really discriminate between the two expressions for science *episteme* and *techne.* (It was just in the special context of the *Ethics* that Aristotle discriminates between these two forms of knowledge: technical reasoning and mathematical demonstration.) Normally Aristotle followed the ordinary language and a special academic language in calling this enterprise of moral philosophy too *episteme* or *techne.* But it is very easy to see the problem. We have *techne* in the narrow sense of the word in what we call craftsmanship and all forms of expertise, and *techne* or *episteme* is introduced in the practical philosophy in terms of an architecture of this system of sciences in which the highest science is politics because it encompasses the whole subordinated professions and forms of knowing. So this well-known system of sciences in the ancient sense, which was already developed in the Platonic dialogues, continues in the Aristotelian description. But he raises a new question: what does it mean that what is found in learning, in craftsmanship or in mathematics, in strategy or in writing is subordinated to politics? Is politics just an expertise of certain technicians of human life and is there a way to teach virtue and to teach in the field of political decision-making—to teach in the sense of conveying a certain knowledge, the truth, to which the pupil can refer as something reliable? Obviously not. Well, to this extent "politics" as moral philosophy can not be a *techne* and teach a set of rules, for to do so would overlook the function of *phronesis* which is just the application of more or less vague ideals of virtues and attitudes to the concrete demand of the situation. Moreover, this application can not evolve by mere rules but is something which must be done by the reasoning man himself.

That is my point: This application, this concretization of the general is the universal aspect of hermeneutics. To understand *in concreto* what the text in general says, that is the task of the jurist in applying law, that is the task of the teacher in explaining the message of the Bible; and for it one needs "prudence." In whatever connection, the application of rules can never be done by rules. In this we have just one alternative, to do it correctly or to be stupid. That is that!

Returning to our problem, we must try to see more precisely what "practical philosophy" entails. We appear to have an impasse. On the one hand we have the clear evidence that practical decision-making and reasoning is not dependant merely on generalities but must be concretized by our practical reasoning and application. On the other hand, there is the game of theoretical description. Aristotle insists that it has a subordinate function. He tells us that it is like the man who tries to hit the goal as an archer, and Aristotle compares his own function with this man. When one tries to hit a goal one tries to concentrate on a special target, like this nail in the wall, just as every hunter when he hunts a deer tries to hit a special point. The hunter and the archer concentrate on a little piece of the whole. Aristotle would say that there is this nail in the wall, this target, so that it is easier to hit the center; that is not, however the full art of this sport but just an addition so as to make it easier.

And so, with respect to hermeneutics and the humanities as a whole we have the task of subordinating both our scientific contribution to the cultural heritage and academic education to a more fundamental project of letting the tradition speak to us. The best definition for hermeneutics is: to let what is alienated by the character of the written word or by the character of being distantiated by cultural or historical distances speak again. This is hermeneutics: to let what seems to be far and alienated speak again. But in all the effort to bring the far near that we make by methodical investigation, in all that we learn and do in the humanities, we should never forget that the ultimate justification or end is to bring it near so that it speaks in a new voice. Moreover, it should speak not only in a new voice but in a *clearer* voice. I would say that the best interpretation should be defined by the fact that one is able to forget it afterwards in the way that one reads the interpreted texts with a certain feeling of self-evidence so that one can resume their own interests, share their own problems, and ask their own questions in order to find a better solution. That, of course, does not happen very often, but it is the hermeneutical ambition. At any rate the convincing interpretation is one which acquires the quality of self-evidence.

With respect to the practical philosophy of Aristotle, it is important to underline one of the qualities of theory, namely, that a certain distantiation is helpful for overcoming a too one-sided commitment of the individual subject. But on the other hand, I think Aristotle was well aware—and in my eyes this is the most important point we have to learn from him and by this re-connection to the older tradition of the philosophy of science—that the preconditions for theorizing in such fields are not neutral objectifications, but articulations of pre-given and lived patterns of social life. So he describes very well what *Verstehen,* understanding, for example, means in practical situations. It is just the privilege of friends that they can give advice; to give advice involves prudence, it means to realize the full moral situation of the other. That is what we call empathy or sympathy, something which is necessary for understanding under a moral point of view. Without it our effort of understanding would not, for example, see the conditions of life by which somebody failed and thus could not offer genuine advice.

This is one of the points I wanted to add to the general description of the problem of the humanities we are dealing with today. I would call it solidarity. When I say that just friends are able to give good advice, I take this to mean that the concept of friendship should be expanded as much as the concept of communicative understanding. And what does a society mean without patterns of self-evident solidarity between human beings, neighbors, members of a family, colleagues in a profession—in every case, a common basis of solidarity? I would say that Aristotle was well aware of this point for any theoretical and conceptual work in ethics and politics. Politics, of course, means not only practice, e.g. of political action or political debate; but there is also the study of the institutions—the well-known content of a tremendous research work of Aristotle. Nevertheless, it remains indubitable for Aristotle that the point is not to enrich theoretical insight for its own sake but to apply this knowledge in a reasonable way in the given circumstances of a given situation. That is the work of political prudence alone.

So, I come back to my initial set of problems. I think we must try to realize how these special conditions which Aristotle works out correspond to the forms in which our reasoning is embedded in our own lines of life, of past, of historical memory and so on. How would Aristotle react to the way in which we put forth new proposals for our political problems? Certainly, he would agree with us that the professional constitution makers are not the right men for developing our practical reason and the rationality of our social behavior. We have to learn from our own needs and from the practice of our own life how to find generalities and to make institutions which promote what is best.

These issues point toward the possibility of better self-understanding in the humanities, an understanding that it is not a matter of mastering matters by information but of trying to participate in our social life and in the heritage of our culture. I heard one day that a Japanese scholar, who was a great admirer of Heidegger, was disappointed because in reading my book he felt excluded, since I insisted so much on the basic grounds of practice and of living in our own tradition. But I think he was wrong. Tradition is not a privilege of Western culture; the dialogue between tradition as well as the dialogue between our past and our tradition is expanding beyond any pregiven limitations. As other cultures are now leaning towards our culture, I think the dialogue or the communicative understanding between them acquires more and more the common ground of human life that is needed for comprehensive partnership and for the general interests of mankind in economics, in energy problems, and in the problems of preservation from catastrophic political developments. In all these things, we no longer pretend any form of exclusiveness and share with everybody our human task of contributing in the modest proportions which are ours to the well-being of human life.

Redaction by James Risser

NOTES

[1]Akademie Edition, XX, 44.